Concise Textbook of Small Animal Handling

Concise Textbook of Small Animal Handling

A Practical Handbook

C. B. Chastain
Emeritus/Adjunct Professor
College of Veterinary Medicine
University of Missouri, Columbia, MO

Line Drawings by Lynn Vellios

CRC Press
Taylor & Francis Group
Boca Raton London

CRC Press is an imprint of the
Taylor & Francis Group, an **informa** business

First edition published 2022
by CRC Press
6000 Broken Sound Parkway NW, Suite 300, Boca Raton, FL 33487–2742

and by CRC Press
4 Park Square, Milton Park, Abingdon, Oxon, OX14 4RN

CRC Press is an imprint of Taylor & Francis Group, LLC

© 2022 C. B. Chastain

Library of Congress Cataloging-in-Publication Data
Names: Chastain, C. B., author.
Title: Concise textbook of small animal handling : a practical handbook / C.B. Chastain.
Description: First edition. | Boca Raton : CRC Press, 2022. | Includes bibliographical
    references and index.
Identifiers: LCCN 2021042746 (print) | LCCN 2021042747 (ebook)
Subjects: LCSH: Animal handling. | Pet medicine.
Classification: LCC SF760.A54 C45 2022 (print) | LCC SF760.A54 (ebook) |
    DDC 636.08/32—dc23/eng/20211018
LC record available at https://lccn.loc.gov/2021042746
LC ebook record available at https://lccn.loc.gov/2021042747

ISBN: 978-0-367-62814-7 (hbk)
ISBN: 978-0-367-62813-0 (pbk)
ISBN: 978-1-003-11092-7 (ebk)

DOI: 10.1201/9781003110927

Typeset in Minion
by Apex CoVantage, LLC

Access the companion website: www.routledge.com/cw/chastain

CONTENTS

PREFACE

Proper handling and restraint are essential to the welfare of captive animals. Animals that are properly handled and restrained can be examined, groomed, and treated in ways that contribute to their optimum quantity and quality of life.

Veterinary medicine is both art and science. Teaching the science of animal handling is relatively easy. Learning the art is more difficult, as it requires much experience, continual practice, and an open mind to new challenges. Each animal is an individual, and each handling environment provides its own advantages and disadvantages.

The most basic part of the art of veterinary medicine is the safe handling of animals. The needs of preveterinary and veterinary students to become knowledgeable in safe, humane animal handling was the impetus to write the CRC textbook *Animal Handling and Physical Restraint*, which is the predecessor to this handbook.

The goals of this handbook are to assist future veterinarians; veterinary technologists, technicians, assistants, and others who deal with animals to be able to handle animals more safely and humanely. It is for quick reference and bedside use. No handling or restraint of animals is without risk, but proper animal handling and restraint aid in reducing the chance of the handler or animal experiencing physical injury or infectious diseases that can be transmitted between species.

Means of confinement (cages, runs, crates, terrariums, aviaries, and environmental enrichments) can affect animal handling, but they are not covered in this handbook. For more information on confinements, see the textbook *Animal Handling and Physical Restraint*.

Each time a handler handles or restrains an animal, he is training it how to accept the next time it is handled. Ideally, the response to the next time it is handled is better than the last.

Those that teach us the MOST about humanity aren't always human. D. L. Hicks, 2015

SUPPLEMENTAL MATERIAL

This book is accompanied by Multiple Choice Self-Assessment Questions, which can be accessed online at www.routledge.com/cw/chastain

ACKNOWLEDGMENTS

I owe many thanks to my mentors on small animal handling, including veterinarians Leonard E. Palmer, Toney Reynolds, Tom Eagle, Cynthia Besch-Williford, and Scott Korte. Other animal handlers I worked with to whom I owe gratitude are Karen Lucy and David Doyle.

Ultimately, the greatest instructors I have had are the animals that I have handled and restrained, who taught me more about safe, humane, and efficient animal handling each time I had the privilege of working with them.

Special thanks go to Lynn Vellios, whose line drawings have permitted nondistracting backgrounds and emphasis on key visual aspects of animal handling. In addition, thanks to Alice Oven, Senior Editor Life Sciences & Veterinary Medicine at CRC Press, Taylor & Francis Group, who has always been encouraging and immensely helpful in the creation of this handbook. The excellent proofreading assistance of Kaitlin Sulkowski is also greatly appreciated.

The encouragement of my wife, Joyce, and daughters, Andrea and Danielle, and my daughters' families as well as their willingness to forgive my obsession with better animal handling has been essential and much appreciated.

C. B. Chastain
July 2021

CAUTION

No handling nor restraint of animals is without risk, but proper animal handling and restraint aids in reducing the chance of the handler or animal experiencing physical injury or transmitting infectious diseases between species.

Learning, acquiring, and maintaining the skill of animal handling and restraint is a methodical progression of stages to (1) learn, (2) see, (3) practice, (4) do, and (5) maintain. To translate knowledge of the art of animal handling into practical skills requires long practice of proper procedures.

Serious injury or death can result from handling and restraining some animals. Safe and effective handling and restraint requires experience and continual practice. Acquisition of the needed skills should be under the supervision of an experienced animal handler.

AUTHOR BIOGRAPHY

 **C. B. Chastain** is a professor emeritus and adjunct professor of veterinary medicine and former Section Head of Small Animal Medicine and Associate Dean for Academic Affairs at the University of Missouri. He also has been on faculty at Iowa State University and Louisiana State University. In addition to being a Diplomate of the American College of Veterinary Internal Medicine, he has authored journal articles; book chapters in veterinary, medical, and allied publications; and textbooks on clinical endocrinology and animal handling and physical restraint. He was also chief editor of the monthly journal, *Advances in Small Animal Medicine and Surgery*, for 20 years.

Prior to entering academic veterinary medicine, he was a horse wrangler for a horseback-riding stable and a licensed guide in Rocky Mountain National Park, Colorado; worked in mixed animal veterinary medical practices in Missouri, Illinois, and New Mexico; and a captain in the U.S. Air Force Veterinary Corps. He has handled, restrained, and trained a variety of animals in differing environments and taught aspects of animal handling for more than 40 years at the undergraduate and professional school levels. Currently, he is a professor and consultant for Veterinary Online Programs at the University of Missouri.

1

SAFER ANIMAL HANDLING AND PHYSICAL RESTRAINT

DOI: 10.1201/9781003110927-1

The reasons to handle or restrain individual animals include physical examination; prophylactic, medical, or surgical treatments; grooming; training; recreation; and companionship.

The single action that veterinary medical personnel do for each and every patient is the application of handling techniques.

HANDLING AND ANIMAL WELFARE

The American Veterinary Medical Association has defined animal welfare as when an animal is healthy, comfortable, well nourished, safe, able to express innate behavior, and not suffering from unpleasant states such as pain, fear, and distress. The Five Basic Freedoms of Domesticated Animals are an international guide for attaining proper animal welfare (Table 1.1).

THE MORE EFFECTIVE HANDLER: THE ART OF FIRM KINDNESS

Proper animal handling for husbandry, treatment, and safety is quiet, methodical, and should leave the animal easier to handle the next time. Guidelines for physical restraint of animals are contained in the American Veterinary Medical Association's position statement on the APhysical Restraint of Animals. The essentials of the AVMA's position statement are listed in Table 1.2.

AFFECTION FOR THE ANIMALS
A good animal handler has to like the type of animal that will be handled. Using food rewards, petting, scratching, grooming, or verbal praises are beneficial under the correct circumstances.

PROPER ATTITUDE
Good animal handlers are calm, deliberate, patient, organized, and determined and they attempt to prepare the animal to better tolerate future handling experiences.

- **Calm, Deliberate, and Patient:** Extroverted behavior, i.e., direct stares, exaggerated facial expressions, frequent hand and arm movements, and loud or spiking speech patterns, can attract the attention of humans, but these mannerisms do not gain or maintain trust from animals. Animal handlers should move and act calmly, deliberately, and patiently.

Table 1.1 Five Basic Needs (Freedoms) of Domesticated Animals	
•	A suitable environment
•	A suitable diet
•	The ability to exhibit normal behavior
•	The need for an animal to be housed with or apart from other animals
•	Protection from pain, suffering, injury, and disease

Table 1.2 Essentials of Proper Animal Restraint	
•	The least restraint required to allow the procedure(s) to be performed properly
•	Protection of both the animal and personnel from harm
•	To plan, formulate, and communicate restraint prior to its application
•	The use chemical restraint when physical restraint presents excessive risk of injury

- **Organized Approach:** Being organized and having a plan before handling or restraining animals is important for success for each handling event and all future handling events with the animals handled.
- **Determination:** Determination is an essential quality of a good handler. A handler must be confident and determined that the plan to handle or restrain an animal will be successful.
- **Controlled Release:** The release of a restrained animal must be as quiet and calm as possible, and it must be under control of the handler, not the animal. Each handling, and especially restraint experience, is a lesson learned by the animal, and their release is most remembered.

ALLOCATION OF SUFFICIENT TIME: POWER OF PATIENCE

Sufficient time to observe the animal or animals to be handled is important in determining the best approach to handling and to allow the animal or animals to adapt to handler presence.

- **Disadvantages of Being in a Rush:** Trying to rush proper handling leads to increased risk of injuries or excessive procedure time because of animals being stressed and scared.
- **More Time for Special Cases:** Longer times need to be allocated for certain animals, particularly for handling young, elderly, sick, or newly acquired animals.

USE OF VOICE, TOUCH, AND BODY LANGUAGE

Restraint begins with the handler's voice or body language.

- **Voice:** Animals like to hear a handler's voice if it is soothing and has rhythmic tones.
- **Body Language:** Animals use body language to a greater extent than any other means of communication.
  - **Lowered Posture:** Handlers greeting small animals are better received if the handler lowers his posture.
  - **Avoiding a Direct Stare:** A glancing gaze or indirect stare by another animal or a handler is less threatening that a direct stare. Staring particularly at an animal's eyes is dominance-challenging to carnivores.
- **Touch:** Touch is used last but can readily convey handler confidence and intentions to an animal being handled. Animals tolerate touch on the shoulders more easily than touch around the more vulnerable areas such as the eyes, ears, throat, belly, or legs.

ALWAYS ON GUARD: SAFETY FIRST

Handlers must be constantly aware of the risks of injury to themselves, other people, or the animal being handled.

- **Individual Animal Reactions Vary:** Factors which affect an individual animal's reactions to handling are familial tendencies; prior handling and training; trust, or the lack of it, in the handler; and stressful events preceding or during the handling. As a result, the assumption that all domestic species or animals within a species are the same can lead to serious injury.
- **Overconfident and Ill Prepared:** Most injuries from animals are caused by the handler being overconfident and undercompetent. Animal handlers must always position themselves and take other precautions to eliminate or minimize the chance of injury to all involved.
- **Special Risks to Veterinary Staff:** Veterinary staff are at a relatively high risk of injury. Many animals they must treat have not been socialized to humans or previously handled. In addition, sick or injured animals may act atypically because of pain or fear and often hide their disease or injury until a handler disturbs them.

DISTRACTION VERSUS PAIN FOR RESTRAINT

Distractions are the basis for most humane and effective animal handling techniques and are not painful when correctly used.

- **Pain:** Pain is a message sent to the brain that body tissue is being injured. Painful procedures leave a physical or psychological mark on the animal.
- **Distractions:** A distraction is applying a nonpainful stimulus that supersedes competing stimuli. Petting is the simplest form of distraction for handling.

RESPECT FOR HANDLERS

Animals that are either fearful of handlers or have no respect for humans are the most dangerous to handle.

- **Gaining Respect:** Leaders of animals establish their leadership by the control of movement and access to resources. Effective handlers do the same. Respect is gained by their knowledge that either pleasure (praise, food treats) or discomfort (not pain) will consistently occur with certain behaviors.
- **Avoiding Fear:** Fear can result from the expectation of pain. If fear is from instinct, it can be moderated. If it is from having experienced pain, it is often permanent.

ADAPTATION TO SPECIAL CIRCUMSTANCES

Animal handling is not a set recipe that fits all situations. An effective animal handler must adapt techniques to the species, the surroundings, and the individual.

- **Surroundings:** Animals may act differently in different surroundings due to noise, lighting, odors, visible distractions, and associations of actual or similar surroundings with past experiences.
- **Current Attitude and Condition:** Each animal handled should first be observed to assess its current attitude and physical condition.
- **Need for Special Handling:** Young, elderly, pregnant, and sick animals always need special handling.

APPROPRIATE ATTIRE, GROOMING, AND PERSONAL HABITS

Proper handler attire for the type of animal handling to be done is important for handlers and animals. Inappropriate attire can be dangerous, such as dandling jewelry or bulky rings (Figure 1.1).

- **Clothing:**
    - **Coat and Footwear:** Clothing should be reasonably clean and untorn. A clinical or laboratory coat and water-impermeable shoes are appropriate for handling dogs, cats, and other small animals (Figure 1.2). Long-sleeved coats can aid in protecting handlers from cat scratches.
    - **Appropriate Time:** Attire for animal handling should be worn only when handling or restraining animals and then changed to reduce the risk of transmitting disease among other animals and to humans.
- **Fingernails:** Fingernails should not extend beyond the end of the finger to reduce the risk of injury to other handlers or to animals being handled and because longer fingernails are more capable of entrapping disease agents.
- **ID Badges:** If ID badges are needed, they should either be attached to the clothing or worn using a safety, breakaway lanyard around the neck.
- **Cuts and Abrasions:** Handler cuts or abrasions should be treated and covered before handling animals.
- **Food and Tobacco Products:** Smoking or consuming food or drink while working with animals or in animal handling areas should be strictly avoided due to the danger of introducing infectious organisms to the handler's mouth.

Figure 1.1 Inappropriate small animal handling attire.

Figure 1.2 Appropriate small animal handling attire.

- **Personal Protective Equipment:**
  - **Waterproof Apron:** Waterproof aprons should be worn when bathing animals.
  - **Eye Protection:** Safety glasses or goggles should be worn if handling animals whose nails or beaks are being trimmed, especially when using a high-speed, handheld electric grinder (Dremel).
  - **Face Protection:** Face masks or shields must be worn when handling birds with pointed beaks or assisting with teeth cleaning or other dental procedures.
  - **Gloves:** Latex rubber or nitrile gloves should be worn if hands have cuts or cracks and should always be worn when handling small mammals other than dogs and cats to protect from infectious diseases and allergens.

CONDITIONS FOR HANDLING AND RESTRAINT

Outcomes of handling and restraint of animals can be affected by health of the animal, the time of day, lighting, ambient temperature, setting and facilities, personnel, and duration.

PREHANDLING CONSIDERATIONS

An unhealthy animal can have an altered temperament requiring special handling techniques or elevated risk of injury from being handled due to illness or previous injury. Before handling any animal, it and its surroundings should be visually inspected (Table 1.3).

- **Signs of Injury or Disease:** The animal should be observed for signs of possible injury or disease, which could alter the means appropriate for handling the animal. Knowledge of the animal's normal posture, movement, and activities, including how and how often it lies down, is important.
- **Vocalizations:** The animals' normal vocalizations should also be known.
- **Appetite, Thirst, Eliminations:** The area of containment (crate, cage, run, etc.) should be examined for urine and feces. Handlers should be familiar with the normal amount, consistency, odor, and color of feces for the type of animal to be handled. If food or water has been present, evidence of whether the animal has been eating or drinking should be investigated
- **Animal Attitude:** The animal's attitude should be observed for signs of depression or aggressiveness.
- **Mobility:** Its ambulation and/or other movements should be screened for lameness or other impaired movements.
- **Respirations:** The respiration rate and depth should be monitored. Impaired respiration can make any handling method particularly dangerous to the animal.

Table 1.3	Prehandling Evaluation of the Animal and Its Surroundings
•	Check for signs of injury or disease.
•	Be alert for abnormal vocalizations.
•	Ascertain current appetite, thirst, and quantity and appearance of eliminations.
•	Observe for signs of abnormal attitude.
•	Assess signs of mobility.
•	Watch for abnormal depth or rate of respirations.

PRERESTRAINT CONSIDERATIONS

Restraint, if needed, must be applied effectively on first attempt, or the animal will learn to escape the restraint in the future.

- **Formulate a Plan**
  - **Check Equipment:** A plan needs to first include a check of equipment.
  - **Capable Assistance:** If others will assist, everyone must be physically capable and trained to handle animals, and they must know the current plan thoroughly.
- **Determine What Safety Precautions Are Appropriate**.
  - **Contingency Plans:** Although a plan for restraint should be designed to be successful on first attempt, a contingency plan should be formulated in case circumstances are unexpected and inappropriate for the initial plan.
  - **Chemical Restraint Option:** Chemical restraint should be ready for some animals in advance of attempted physical restraint so that if chemical restraint becomes necessary, it can be administered without delay.

EFFECTS ON ANIMALS

Past handling experiences can be an asset or impediment to current handling success.

- **Current and Future Effects:** To yield the best lesson that might be learned by the animal, a handler should use the minimum needed restraint, maintain a calm environment, and carefully manage the final release to be perceived that the release was the handler's choice, not the animal's.
- **Impact of Past Experiences:** If adverse effects occur when handling or restraining an animal, the animal will associate events, people, and objects that immediately preceded the handling and occurred during the animal's handling. They may respond with signs of stress or fear when exposed to similar handler clothing, locations (cage, pens, etc.), and other sounds on reexposure.

SURROUNDINGS AND CONDITIONS

The time of day or amount of light can affect animal handling.

- **Ambient Lighting:** Nocturnal (night active) animals are more docile when handled in bright light. Diurnal (daytime active) animals are more docile when handled in subdued light.
- **Reduce Risk of Injury or Escape:** When handling small animals, room doors and windows should be closed and all countertops cleared.

PERSONNEL

All persons who handle animals should be willing, knowledgeable, capable, and able to assist in proper handling and restraint.

- **Qualified Assistants:** It is the primary handler's responsibility to make sure that any assistant is knowledgeable in proper animal handling and is fully aware of the handling or restraint plan. An assistant should also be behaviorally mature and physically strong enough to carry out the needed assistance.
- **Unqualified Assistant Liability:** When veterinarians or veterinary technicians are the primary handlers, animal owners should not assist with handling and restraint of their own animal. Otherwise, the veterinary personnel are liable for resulting owner injury.

DURATION

The duration of animal handling or restraint should be as short as possible to complete the task. Longer durations cause unnecessary stress to the animal and exhausts their patience to tolerate the handling. Prehandling preparation is absolutely essential to minimize the duration of handling and restraint.

GENERAL RISKS

Few domestic animals are naturally aggressive toward humans. When fearful or stressed, most animals' first reaction is to attempt to flee. When fleeing is not an option, they will resort to their means of offense or defense. Causes of handler injuries are listed in Table 1.4.

- **Dog Bites:** Five percent of all emergency-room visits are caused by dog bites. Dogs that receive inadequate early socialization with humans and continued gentle handling, as well as those that are tethered for long periods, are the most likely to bite.
- **Cat Bites:** Bites from cats are common but less life-threatening than most dog bites. The risk of infection from cat bites is greater, and impairment of hand function is a significant risk.

RISKS TO VETERINARY PERSONNEL

Based on the U.S. Bureau of Labor Statistics, veterinary medicine is among the top 10 most dangerous jobs. More than half of all veterinarians and nearly all veterinary technicians will be seriously injured by animals some time in their careers.

- **Location:** The body location of more than half of bites and scratches are to the hands.
- **Cat Bites versus Dog Bites:** Cat bites to veterinarians occur 50% more than dog bites, and they are three times more common in veterinary technicians than dog bites. Scratches from cats are more common than bites but less frequently a reason for seeking medical attention.

RISKS OF DISEASE TO HANDLERS AND OTHER ANIMALS

ZOONOSES: TRANSMISSION OF DISEASE FROM ANIMALS TO HUMANS

A zoonosis is any infectious disease of animals that can be transmitted to humans under natural conditions.

- **Incidence:** Of the more than 1,400 known infectious diseases of humans, 60% are zoonotic. Most are associated with the gastrointestinal tract and transmission is fecal-oral.
  - **In the U.S.:** More than 50 zoonotic diseases are known to be present in the U.S.
  - **U.S. Zoonotic Disease Examples:** These include rabies, salmonellosis, ringworm, plague, tularemia, psittacosis, among others.
- **Means of Transmission:** Transmission of zoonotic disease can be by contact, including bites or scratches; oral or ocular, aerosol (inhaled), or vector borne.

Table 1.4 Causes of Handler Injury	
•	Lack of animal handling knowledge
•	Overconfidence or underconfidence
•	Begin rushed
•	Becoming angered
•	Error by an assistant
•	Pain experienced by the animal
•	Equipment failure

- **Risks:**
    - **Realistic Perspective:** Every animal can carry some diseases that humans could acquire. However, handling apparently healthy domestic animals using basic sanitary practices, such as keeping hands away from eyes, nose, and mouth, keeping skin cuts covered, and washing hands after handling animals carries very little risk of acquiring zoonotic disease.
    - **Higher-Risk Situations:** High-risk animals for transmitting zoonoses are the young, females giving birth, and unvaccinated, stray, or feral animals. Others include those fed raw-meat diets, kept in crowded conditions, and with internal or external parasites. In addition, all reptiles and wild or exotic species are high-risk sources of zoonotic diseases.
    - **Role of Stress:** Stressful handling, including prolonged transportation, or overcrowding of animals can increase the risk of animals shedding disease organisms.
- **Immune Status:** The risk of disease transmitted from animals is greater among people with immature, declining, or impaired immune systems. Young children and immunosuppressed adults should especially avoid nursing calves, all reptiles, and baby chicks and ducklings.
    - **Young Children:** Children 5 years old or younger should have supervised exposure to animals due to immature immune systems and a tendency to put unwashed hands in their mouths.
    - **Advanced Age:** Animal handlers that are more than 70 years old may have increased risk of zoonoses from declining immune responses.
    - **Diseases, Pregnancy, and Medications:** Some conditions, diseases, or treatments in humans, regardless of their age, may lower their resistance to zoonoses.
        - **Diseases:** Diseases that suppress immunity include systemic diseases such as HIV, congenital immunodeficiencies, diabetes mellitus, chronic renal failure, alcoholism, liver cirrhosis, malnutrition, and certain cancers.
        - **Pregnancy:** Pregnancy may also reduce the nonpregnant immune response.
        - **Medications and Other Treatments:** Treatments for cancer, organ or bone marrow transplants, and autoimmune diseases that can depress immunity. Splenectomy and long-term hemodialysis are also treatments that can suppress immunity.
- **Prevention of Transmission:**
    - **Maintain Health:** Keeping animals healthy can also lower the risk of zoonosis and transmission to humans. Routine veterinary care, vaccinations, and parasite screenings should be maintained.
    - **Maintain Healthy Nutrition:** High-quality food is advisable. Dog, cat, or ferret foods that contain any supplementary egg, poultry, or meat products should have been adequately cooked. Raw pet foods can be sources of zoonotic bacteria, such as *Salmonella*.
    - **Avoid Exposure:** Pets should be prevented from drinking from toilet bowls or eating garbage, hunting wildlife, or eating other animals' feces. All pets should be kept away from areas where human food is prepared.
- **Handler Sanitation and Health Precautions:**
    - **Hand Washing:** Hand washing is essential to controlling the transmission of disease (*Procedural Steps 1.1*). Alcohol-based rubs are effective against most disease-producing agents if the hands are not visibly soiled with organic material.

Procedural Steps 1.1	Hand Washing Procedure
1.	Clean fingernails and remove rings.
2.	Wet hands.
3.	Apply an olive-size amount of liquid soap to a palm.
4.	Scrub both hands while counting to 20 slowly.
5.	Rinse thoroughly and dry with paper towels.

- **Wound Care:** If bitten or scratched, the wound should be thoroughly cleaned with warm, soapy water, compression should be applied if bleeding persists, and a physician should be consulted.
- **Sick Animals:** Special precautions are needed if working with animals with diarrhea or skin or mouth sores.
- **Pregnancy:** Pregnant women should not handle cat litter or ewes in the process of lambing.
- **Vaccination Protection:** All animal handlers should be vaccinated against tetanus every 10 years, as recommended by the U.S. Centers for Disease Control. Veterinary personnel are also advised to receive preexposure vaccination against rabies and have serum titers checked every 2 years.
- **Control of Disease Vectors:**
  - **Ectoparasites:** Some animal-related diseases are transmitted to humans indirectly via ectoparasite vectors, such as mosquitoes (encephalitis viruses), ticks (Rocky Mountain Spotted Fever and many others), and fleas (Cat Scratch Disease). The animal carrying the ectoparasite may or may not become ill. Ectoparasites are controlled in dogs and cats with individually applied topical insecticides and acaricides.
  - **Rodents and Birds:** Rodents and birds can also be disease vectors. These are controlled by eliminating entry to animal dwellings and hiding places. Access to food sources should be eliminated by maintaining food storage in rodent-proof sealed containers and proper disposal of garbage.
- **Increased Risks for Veterinary Personnel:**
  - **Greatest Risks:** The greatest zoonotic risks to companion animal veterinarians have been reported to be ringworm, rabies, and antibiotic-resistant bacterial infections.
  - **Direct Transmission:** Direct transmission can be contact with animal saliva, blood, urine, or feces with handler eyes, nose, or mouth, which can occur from splashing of body fluids or eating, smoking, or touching the face. Contamination of a skin cut, scratch, or crack is also a form of direct transmission.
  - **Indirect Transmission:** Vector-borne indirect transmission can be the bite of a fly, mosquito, tick, or flea carrying a zoonotic organism.
- **Personal Protective Equipment:** Personal protective equipment (PPE) should be considered in possible zoonotic risk situations.
  - **Eyes and Face:** PPE can include protection of the eyes with properly fitted goggles or ANSI-approved face masks. Ears should be protected from excessive noise with earmuffs or molded ear plugs (cotton plugs are insufficient).
  - **Torso:** Protection for the torso can be lab coats, coveralls, gowns, or aprons.
  - **Arms and Hands:** Long sleeves protect arms against scratches and splashes. Hands are typically protected with rubber or nitrile gloves.
  - **Scalp:** The scalp can be partially protected from exposure to cuts, splashes of infectious liquids, and ringworm with a hat.
  - **Feet:** Feet may be protected with closed-toe, slip-resistant, water-impermeable shoes or boots.
- **Summary of Recommendations to Prevent Transmission of Zoonotic Disease:**
  - **Essential Preventive Measures:** Essential recommendations for the prevention of zoonotic disease are listed in Table 1.5.
  - **Additional Information for Veterinary Personnel:** For precautions for veterinary personnel handling overtly sick animals, consult the current *Compendium of Veterinary Standard Precautions for Zoonotic Disease Prevention in Veterinary Personnel*, published annually in the *Journal of the American Veterinary Medical Association* and *Veterinary Standard Precautions* at www.nasphv.org/.

Table 1.5 Recommendations to Prevent Zoonotic Diseases

•	Thoroughly wash your hands after feeding or touching animals or moving their waste; do not dry hands on clothing.
•	Do not permit animals to eat from human plates or utensils.
•	Keep pets supervised to prevent hunting at will, and do not feed raw meat.
•	Do not eat or drink in animal handling areas.
•	Wear appropriate clothing when handling animals.
•	Do not kiss animals.
•	Wash cuts thoroughly
•	Wear gloves when gardening.
•	Keep animal environment reasonably clean, and prevent children from playing where there is animal waste.
•	Clean cat litter daily and wash your hands immediately afterward.
•	Keep animals from household areas where human food is prepared or handled.
•	Do not bathe pets in sinks or bathtubs used by humans.
•	Deworm animals on a regular basis and provide reasonable control of fleas, ticks, and mosquitoes.
•	Avoid stray animals.
•	Do not keep wild animals as pets.
•	Vaccinate animals against zoonotic diseases and maintain tetanus vaccinations in all animal handlers and rabies vaccinations in high-risk animal handlers.
•	Use proper low-stress handling techniques and containment practices and facilities to reduce stress-induced shedding of zoonotic diseases.
•	Routinely train animal handlers on the prevention of zoonotic disease and animal handling safety measures.

TRANSMISSION OF DISEASE AMONG ANIMALS BY THEIR HANDLERS

Handlers can transmit diseases among animals if proper sanitation and disinfection are not practiced.

- **Sanitation and Disinfection:** *Sanitation* (reduction of possible disease agents) or *disinfection* (complete or nearly complete elimination of disease agents) techniques are needed to reduce the chance of inanimate objects (clothing, handling equipment, confinement structures) from becoming inanimate transmitters of disease (*fomites*).
    - **Cleansing:** Disinfection should be preceded by cleaning of all organic matter (feces, blood, saliva, dust, urine, and hair) before using the disinfectant.
    - **Careful Preparation of Disinfectants**
        - **Dilution:** The manufacturer's directions for dilution and use of the disinfectant should be closely followed.
        - **Practical Disinfection:** A common, effective, and inexpensive disinfectant (sterilant) is household bleach diluted to 1:32 (1 cup bleach per gallon of water). **NOTE:** Bleach (sodium hypochlorite) must never be mixed with an acid or ammonia, which will result in the release of toxic gases.
- **Risk of Transmission (*Procedural Steps 1.2*):**
    - **Closed Group:** The risk of transmission of disease is low if all animals appear healthy and belong to the same household. When animals drink from the same water source, eat from the same ground or containers, touch noses, and have other frequent physical contacts and appear healthy, the risk of handling procedures spreading disease is mild to nonexistent.
    - **Segregated Age Groups:** The risk is lowered further if animals of different age groups are segregated. Older animals are more likely to be disease carriers without signs and capable of transmitting disease to younger animals.

- **Separation of Sick Animals:** Sick animals should always be isolated or segregated and handled separately after disinfection or change of handling tools and clothing. When handling animals from different households, disinfection of handling tools and clothing should take place between handling the different groups of animals.

Procedural Steps 1.2
Order of Handling to Prevent Disease Transmission
First: Handle young, apparently healthy animals.
Second: Handle apparently healthy adult animals.
Finally: Handle isolated or sick animals.

- **Containment**
  - **Wash Hands:** To prevent the transmission of disease, handlers should always wash their hands after handling animals.
  - **Proper Foot Covering:** Water-impermeable boots should be worn if walking on surfaces on which urine or feces may have been present, and boots disinfected before moving to another animal holding area.
  - **Clean Clothes:** Clothes worn during handling of animals that may have been ill should be washed near handling areas with commercial equipment.
  - **Clean Cages and Runs:** Animal confinement areas should be properly cleaned before introducing new animals.
  - **Quarantine New Animals:** New animals being introduced to an established group of animals should be held in quarantine until the effects of transport stress and the typical incubation period for infectious diseases has passed (usually 10 days).

ANTHROPONOSIS: TRANSMISSION OF DISEASE FROM HANDLERS TO ANIMALS

In rare cases, diseased handlers can transmit their infection, such as tuberculosis, to animals. This is referred to as reverse zoonoses or *anthroponosis*. If sick, animal handlers should not handle animals due to the risk of reverse zoonosis, as well as the added risk of physical injury from impaired judgment and delayed reactions.

ETHICAL CONCERNS

Ethics are based in part on social mores and therefore not static. Methods of animal handling, restraint, and discipline once considered acceptable may not be tolerated by society today. Acceptable techniques are changing. Certain actions are considered inappropriate for handling, restraining, or disciplining companion animals (Table 1.6).

Table 1.6	Unacceptable Handling, Restraining, or Disciplining Animals
•	Use of force beyond that needed for self-defense or protection of others
•	Use of force as punishment
•	Punishment delivered in anger or to inflict pain
•	Striking an animal on the head or other sensitive or injured body parts
•	Choking an animal
•	Shaking an animal violently
•	Striking an animal with a rigid object, if not to avert a dangerous attack
•	Unnecessary use of chemical restraint

USE OF FORCE

Force is considered permissible if handlers are in full control of their emotions at the time and only the minimum amount of force needed is used to protect the safety of humans, the animal being handled, or other animals. Force must be used with consideration of the animal's nature and with empathy for the animal.

RESPONSIBILITY FOR THE ACTIONS OF ASSISTANT HANDLERS

Handlers who are also supervisors of other animal handlers bear the responsibility to ensure that the other handlers are appropriately trained and supervised.

- **Written Guidelines:** Written guidelines, although they may only deal with extremes, should be provided along with a no-tolerance policy on cases of animal abuse.
- **Penalties for Abuse:** Immediate termination of employment should be written and understood by all employees handling animals as a consequence to unequivocal animal abuse. It is well established that there is a link between willful abuse to animals and eventual domestic violence against humans.

LEGAL CONSIDERATIONS

Most states have laws that require animal handlers to exercise adequate control over animals to prevent them from harming themselves, other animals, people, or property.

LIABILITY

Legal liability is being responsible for others' safety, animals, and other assets.

- **Waivers:** Liability waivers signed by people involved in animal handling other than the primary handler may help win a case for a defendant, but they do not prevent lawsuits and the cost of defense. Waivers of responsibility also do not absolve handlers of liability in an injury or death that is due to their negligence or incompetence.
- **Incompetence:** Incompetence is simply not having the knowledge or ability to control an animal. It is important to ascertain that any assistant animal handler is mature enough, strong enough, and trained sufficiently for each task to be performed.
- **Negligence:** Failing to properly contain or control an animal that causes injury to a human is negligence. Knowing that an animal is potentially dangerous and not taking extra efforts to protect others is also considered negligence. Inadequately training assistants can be considered negligence on the part of the primary handler (supervisor).
- **Indirect Injuries:** Injury received in an attempt to flee from an animal demonstrating threatening behavior can be owner or agent negligence.
- **Insurance:** It is prudent to have adequate liability insurance that covers activities of an animal handler.

INHERENTLY DANGEROUS ANIMALS

Wild and exotic animals, in addition to being rabies reservoirs, are considered inherently dangerous animals.

- **Wild and Exotic Animals:** Some animals that are inherently dangerous include lions, tigers, cougars, bears, monkeys, venomous snakes, and large constricting snakes more than 8 feet in length.
- **Rabies Reservoirs:** Animals which are considered rabies reservoirs (bats, skunks, coyotes, foxes, and raccoons) should not be handled except when absolutely necessary and only by professionals trained to handle wildlife.

ANIMAL ABUSE REGULATIONS

Domestic animals have traditionally been viewed legally as property. Still, there are laws to protect against the inhumane treatment of animals.

- **Felony Potential:** In more than 30 states, at least one form of animal cruelty constitutes a felony. The Federal Bureau of Investigation (FBI) reclassified crimes against animals in 2016 as a Group A offense and included cases in the FBI's National Incident Based Reporting System.
- **Veterinary Professionals:** If the handler is a veterinary medical professional, there is risk of malpractice charges that could lead to disciplinary action under the state veterinary practice act.
- **Stepping-Stone to Abuse of Humans:** Willful abuse of animals is known to be associated eventually with domestic violence, child abuse, and other violent crimes toward people.
- **Abuse of Animals for Research:**
  - **Protection of Animals in Federally Funded Research:** The most stringent restrictions on the handling of animals involve those used in federally funded research.
  - **Oversight of Federally Funded Research**
    - **OLAW:** The NIH Office of Laboratory Animal Welfare oversees all federally funded research institutions using laboratory animals.
    - **ACUC:** Institutions are required to have Animal Care and Use Committees that monitor the care and use of research animals. A veterinarian must be employed by the institution to oversee the care and use of the institution's research animals and be an advisor to the ACUC.
    - **AAALAC:** The Association for Assessment and Accreditation of Laboratory Animal Care International requires a higher standard of care than the NIH, but membership is voluntary.

ROLES OF CHEMICAL AND PHYSICAL RESTRAINT

PROPER USE OF CHEMICAL RESTRAINT

Chemical restraint should only be used when physical restraint techniques are substantially less safe for the animal or the handler, not just for convenience, to supplement income, or as a substitute for good handling and physical restraint methods.

ADVANTAGES AND DISADVANTAGES OF CHEMICAL RESTRAINT

- **Potential Advantages of Appropriately Used Chemical Restraint:**
  - **Safety:** Recent innovations in chemical restraint (sedation and anesthesia) have been highly beneficial to animals, owners, and veterinarians in alleviating animal stress and possible physical injury.
  - **Convenience:** The convenience of chemical restraint can lead to decreased risk to the handler and shorter handling time.
- **Potential Disadvantages of Unnecessary Chemical Restraint:**
  - **Altered Vital Signs of the Physical Exam:** Chemical restraint can also interfere with a physical exam by altering vital signs (heart rate, respiratory rate, and body temperature).
  - **Adverse Health Effects:** All sedatives and anesthetics have potentials to cause adverse health effects (Table 1.7).
  - **Fostering Fear:** Chemical alteration of consciousness may alleviate some of the fear and resistance to restraint in animals, although in some cases, memory of the loss of full control of their body during induction or recovery may instill fear in other animals.

Table 1.7	Potential Adverse Health Effects of Chemical Restraint
•	Respiratory or cardiac arrest
•	Physical injuries to animals or handlers during the chemical administration, induction, or recovery
•	Organ damage from overdosage, individual variation in response, drug interactions, or idiosyncratic reactions.

ANTIANXIETY DRUGS TO FACILITATE HANDLING OF SMALL ANIMALS

Anxiety is a normal response to a new situation. Learning is facilitated by anxiety, and all anxiety does not need to be suppressed with drugs. Handling an anxious animal without sedation in a nonpainful, quiet, calm way is a learning situation that can help the animals become less stressed in the present and future situations.

- **First-Generation Antihistamines**
  - **Diphenhydramine:** Diphenhydramine is an overt-the-counter antihistamine that can cause mild sedation and relieve anxiety in some dogs.
  - **Chlorpheniramine:** Chlorpheniramine is an over-the-counter antihistamine that can cause mild sedation and relieve anxiety in some cats.
- **Acepromazine:** Acepromazine is a prescription tranquilizer for both dogs and cats. It can moderate anxiety but may cause temporary protrusion of the third eyelid and lower blood pressure. Cats may experience paradoxical excitation.
- **Diazepam, Alprazolam, and Other Benzodiazepines:** Diazepam and alprazolam are Schedule IV, controlled prescription sedative hypnotic antianxiety drugs used off-label in dogs and cats. They may cause ataxia and excessive drooling. In rare cases, anxiety may be increased.
- **Imepitoin:** Imepitoin is an imidazolone that is a partial agonist of the benzodiazepine sites. It is an anticonvulsant with proposed antianxiety effects in dogs and cats. It is approved in the U.S. for use in dogs to treat noise phobias. It can cause ataxia, increased appetite, vomiting, and sleepiness.
- **Trazodone and Fluoxetine:** Trazodone and fluoxetine are serotonin-antagonist/reuptake inhibitor drugs marketed as antidepressants in humans and used off-label to control anxiety in dogs or cats. They may cause dilated pupils, protruding third eyelid, vomiting, diarrhea, panting, ataxia, hypotension, or arrhythmias. In rare cases, anxiety may be increased.
- **Clomipramine:** Clomipramine is a tricyclic antidepressant and a serotonin reuptake inhibitor marketed for use in humans and in dogs for the control of anxiety. It can cause nausea, vomiting, diarrhea, increased appetite, urine retention, and increased intraocular pressures.
- **Gabapentin:** Gabapentin is a prescription drug marketed for the control of seizures in humans and often used off-label as an antianxiety drug for cats. Britain classified gabapentin as a Class C controlled drug in April 2019 due to risks of abuse and addiction. It may cause ataxia, drooling, and vomiting.
- **Alpha-2 Agonist Sedatives**
  - **Clonidine:** Clonidine is a prescription drug marketed for humans with resistant high blood pressure. It is used in dogs for anxieties and phobias. It may cause ataxia, bradycardia, hypotension, and in some cases, increased anxiety and agitation.
  - **Dexmedetomidine:** Dexmedetomidine is a prescription injectable drug marketed for moderate to deep sedation with analgesia for dogs and cats and as an oral gel for noise aversion in dogs. It may cause vomiting and bradycardia. Atipamezole is an alpha-2 antagonist that can reverse the effects of dexmedetomidine.

- **Tiletamine:** Tiletamine is a dissociative anesthetic chemically related to ketamine. It is a Schedule III controlled drug in the United States. Approved as an injectable combined with zolazepam for dogs and cats, it has been advocated for chemical restraint when administered off-label by buccal administration. It can cause drying of the eyes, hypersalivation, tachycardia, hypertension and later decreased cardiac output, tachypnea, hypoxemia, and cyanosis.

CHEMICAL RESTRAINT FOR DEEP SEDATION OR GENERAL ANESTHESIA

Drugs that cause deep sedation or general anesthesia may be used for temporary control of dangerous small animals, but these replace handling techniques; they do not facilitate handling of small animals in a state of consciousness.

KEYS TO GOOD HANDLING OF ALL ANIMALS

Good handling of any type of domestic or tame nondomestic animal involves proper preparation to become an animal handler (*Procedural Steps 1.3*) and 10 basic keys of good animal handling (Table 1.8).

Procedural Steps 1.3	**Proper Preparation to be an Animal Handler**
1.	Read about animal handling.
2.	Observe normal animal behavior.
3.	Gain guidance from a good handler.
4.	Observe a good handler with animals.
5.	Practice under a good handler's supervision.

Table 1.8 Keys to Good Animal Handling
• Frequently but briefly and gently handle young animals during their critical socialization period to reduce their natural fear of humans while being mindful not to eliminate their inherent respect for humans.
• Quietly handle healthy adults frequently for short periods to habituate the animals for handling without an association with fearful, painful events to follow.
• Confine animals in environments as similar to their ancestors' natural habitat as reasonably possible, e.g., crates as dens for dogs, aviaries for finches, and deep substrate as desert sand for hamsters.
• Provide environmental enrichments that will prevent or reduce boredom and stereotypic behaviors.
• Confine animals with as much personal space as needed to prevent or minimize stereotypic behaviors.
• Maintain social like-species support for animals by keeping prey, pack, and flock animals in groups with a size appropriate for the species.
• Minimally handle elderly, neonatal, or sick animals to prevent their exhaustion or pain.
• Handle animals with confidence, using smooth, rhythmic movements along with a calm low-pitched voice.
• Be able to recognize abnormal behavior for the species and the individual, including fear or signs of health problems.
• Use correct timing and type of responses to favorable or unfavorable animal conduct to shape their future behavior to handling.

1. Bender JB, Shulman SA. Reports of zoonotic disease outbreaks associated with animal exhibits and availability of recommendations for preventing zoonotic disease transmission from animals to people in such settings. J Am Vet Med Assoc 2004;224:1105–1109.

2. Centers for Disease Control and Prevention Morbidity and Mortality Weekly Report: Compendium of Measures to Prevent Disease Associated with Animals in Public Settings, 2011;60(RR04):1–24.

3. Engle O, Muller HW, Klee R, et al. Effectiveness of imepitoin for the control of anxiety and fear associated with noise phobia in dogs. J Vet Intern Med 2019;33:2675–2684.

4. Engel O, von Klopmann T, Maiolini A, et al. Imepitoin is well tolerated in healthy and epileptic cats. BMC Vet Res 2017;13:172–179.

5. Epp T, Waldner C. Occupational health hazards in veterinary medicine: Zoonosis and other biological hazards. Can Vet J 2012;53:144–150.

6. Epp T, Waldner C. Occupational health hazards in veterinary medicine: Physical, psychological, and chemical hazards. Can Vet J 2012;53:151–157.

7. Esch KJ, Petersen CA. Transmission and epidemiology of zoonotic protozoal diseases of companion animals. Clin Microbiol Rev 2013;26:58–85.

8. Fowler HN, Holzbauer SM, Smith KE, et al. Survey of occupational hazards in Minnesota veterinary practices in 2012. J Am Vet Med Assoc 2016;248:207–218.

9. Kipperman BS. The role of the veterinary profession in promoting animal welfare. J Am Vet Med Assoc 2015;246:502–504.

10. Landercasper J, Cogbill T, Strutt P, et al. Trauma and the veterinarian. J Trauma 1988;28:1255–1259.

11. Langley, RL, Hunter JL. Occupational fatalities due to animal-related events. Wilderness Environ Med 2001;12:168–174.

12. Lilly ML, Arruda AG, Proudfoot KL, et al. Evaluation of companion animal behavior knowledge among first-year veterinary students before and after an introductory animal behavior course. J Am Vet Med Assoc 2020;256:1153–1163.

13. Lucas M, Day L, Shirangi A, et al. Significant injuries in Australian veterinarians and use of safety precautions. Occup Med 2009;59:327–333.

14. McGreevy P. Firm but gentle: Learning to handle with care. J Vet Med Ed 2007;34:539–541.

15. McMillian M, Dunn JR, Keen JE, et al. Risk behaviors for disease transmission among petting zoo attendees. J Am Vet Med Assoc 2007;231:1036–1038.

16. National Association of State Public Health Veterinarians Animal Contact Compendium Committee. Compendium of measures to prevent disease associated with animals in public settings, 2013. J Am Vet Med Assoc 2013;243:1270–1288.

17. National Association of State Public Health Veterinarians. Compendium of veterinary standard precautions for zoonotic disease prevention in veterinary personnel. J Am Vet Med Assoc 2010;237:1403–1422.

18. Nejamkin P, Cavilla V, Clausse M, et al. Sedative and physiologic effects of tiletamine-zolazepam following buccal administration in cats. J Feline Med Surg 2020;22:108–113.

19. Patronek GJ, Lacroix CA. Developing an ethic for the handling, restraint, and discipline of companion animals in veterinary practice. J Am Vet Med Assoc 2001;218:514–517.

20. Poole AG, Shane SM, Kearney MT, et al. Survey of occupational hazards in companion animal practices. J Am Vet Med Assoc 1998;212:1386–1388.

21. Stevens BJ, Frantz EM, Orlando JM, et al. Efficacy of a single dose of trazodone hydro-chloride given to cats prior to veterinary visits to reduce signs of transport- and exami-nation anxiety. J Am Vet Med Assoc 249:202–207.

22. Van Haaften KA, Forsythe LRE, Stelow EA, et al. Effects of a single preappointment dose of gabapentin on signs of stress in cats during transportation and veterinary examination. J Am Vet Med Assoc 251:1175–1181.

23. Van Soest EM, Fritschi L. Occupational health risks in veterinary nursing: An explor-atory study. Aust Vet J 2004;82:346–350.

24. Westgarth C, Brooke M, Christley RM. How many people have been bitten by dogs? A cross-sectional survey of prevalence, incidence and factors associated with dog bites in a UK community. J Epidemiol Community Health. 2018;72:331–336.

25. Willems RA. Animals in veterinary medical teaching: Compliance and regulatory issues, the US perspective. J Vet Med Ed 2007;34:615–619.

2

SMALL ANIMAL BEHAVIOR

DOI: 10.1201/9781003110927-2

Animals are not simply hairier versions of humans. Reacting to animals as if they are humans is called *anthropomorphism*, and anthropomorphism is not effective in establishing a safe and effective relationship with animals. Better animal handling involves the handler assuming a leadership role, having empathy for animals, and having knowledge of the normal behavior and needs of the species involved.

BEHAVIOR AND HANDLING

An important foundation for proper animal handling is learning the normal behavior for the species.

- **Natural Instincts:** Knowing the natural instincts of a species is essential to being able to handle, move, and contain them humanely with minimal stress and risk of injury to either the handler or the animals.
- **Reading Body Language:** Most communications between animal species is silent, i.e., via *body language*.
  - **Initial Observation:** Behavior and attitude of animals should be assessed by observing them at a distance before an approach for handling. Animals will begin assessing handlers by the handler's body language upon first notice of presence of the handler and modifying their own behavior. Not approaching them immediately will reduce some of the threat they might otherwise perceive.
  - **Avoid Appearance of Submissiveness or Being a Predator:** It is essential to not be perceived by animals as either their predator or being of lesser social rank to them.
  - **Evaluate for Signs of Aggression:** Most animals will telegraph, in advance, their intent to display open aggression. The natural tendency is to avoid, if possible, the possibility of injury and death. Therefore, barking or other vocalizations, lunging, and other indications usually precede an attack on an apparent adversary.
- **Facilitating Handling:** Avoidance of handling resistance in animals requires early socialization with humans and never causing pain or extreme fear (Table 2.1).

PREDATOR OR PREY BEHAVIOR

All domestic animals evolved as either meat eaters (predators) or food for meat eaters (prey). Common small domestic animals, i.e., dogs, cats, and ferrets, are genetically predators. Mice, guinea pigs, and rabbits are prey. Rats can be either, depending on the circumstances.

PREDATORS

- **Vision:** Predators have eyes that are positioned forward in their skulls, which permits greater overlapping of the field of vision from right and left eyes and improves depth of vision. Predators stare directly at their prey and are able to track the movement and judge speed of their prey.

Table 2.1 Causes of Irreparable Handling Resistance in Animals
• Failure to properly socialize animals in their juvenile period of life
• Infliction of pain at any age
• Exposure to extremely fearful situations at any age

- **Pursuit of Prey:** Dogs are pack predators that run after their prey. They are instinctively more aggressive if in a group. They are more aggressive to humans with small stature and if body language is fearful, especially when a human runs away as a prey animal would do.
- **Stalking of Prey:** Cats are solitary hunting, stalking predators. High-pitched sounds and wiggly movements stimulate their predator urges. Most of their stalking is by remaining motionless. Eventually, they have explosive movements used to quickly pounce at their prey from a short distance away.

PREY

- **Vision:** Prey animals have eyes that are located on the sides of their skulls. Prey animals monitor their peripheral environment to detect the presence of predators. They do not stare at predators except to assess their intent and decide on a means of escape.
- **Safety in Numbers:** Most prey animals are gregarious. Living in groups provides more sentinels of danger and sacrificial members for predators. Group-loving prey animals are stressed whenever isolated from their own species.

PERCEPTION OF HANDLERS AS PREDATORS OR PREY

A handler's body language may unintentionally mimic prey or predator behavior.

- **Vision:** Humans have predator eyes, directed forward. Staring directly at dogs or standing over them is considered a challenge for dominance by another predator.
- **Perception as Prey:** Dogs interpret a human running away from them as prey behavior. Playing with cats by wiggling a finger or toe can unintentionally invite a predator attack to a hand or foot.

ANIMAL HIERARCHY: SOCIAL DOMINANCE

Each animal is an individual, the total of a unique combination of genes and past experiences. General assumptions about behavior might be made based on species, age, gender, and breed, but an individual animal may act and react in a unique fashion.

SOCIAL RANK

- **Assessing Social Rank:** Nearly all domestic animals prefer to live in groups. Within animal groups, there is a hierarchy, a social ladder. Knowing the social rank of an individual animal mingling in a group can be helpful in determining the best means of handling or restraint of a particular animal.
- **Change in Social Rank:** The previous status of a group member or that of an animal handler can be altered within a group of animals if the demeanor of the group member or handler is different than usual. Acting ill, injured, or less confident as with advanced age can reduce the status of the individual within a group.

LEADERSHIP

- **Establishing Leadership:** Being dominant to animals requires the control of resources, such as access to food, and movement. Corrections in behavior of others must be clear, metered, consistent, and within seconds of the misbehavior.
- **Demonstration of Leadership:** Dominant status is conveyed primarily by demeanor, particularly calmness and confidence. Force that inflicts pain is reserved for self-defense or defense of the species, such as conflicts during mating seasons. An effective animal handler must be dominant by their demeanor, not by applied force or micromanagement, to animals being handled for the safety of both.

- **Animal Leaders:** Dominance within a group is generally related to height, weight, gender, and age.
  - **Juvenile Play:** Wrestling play during prepubertal life aids in an animal attaining social status as an adult.
  - **Deference:** Dominant animals are identifiable by the deference given them by others, not by displays of aggression and force.
  - **Elevated Position:** Dominant animals will seek a position physically above others. For example, dominant cats will seek ledges, and dominating dogs will attempt to stand over submissive dogs.

SOCIALIZATION WITH HUMANS

Critical periods for socializing animals for less fear toward humans is lengthened by the degree of species domestication and is later and longer in predators than prey animals.

SOCIALIZATION OF DOGS AND CATS

- **Socializing Dogs:** Dogs, the longest-domesticated species, have the longest period of critical socialization with humans which is up to 4 months of age. In contrast, wolves must be handled in the first 14 days of life to have lasting effects on their social behavior with humans.
- **Socializing Cats:** The period for socializing domestic cats to humans is up to 7 weeks of age.

RISKS OF EARLY HANDLING

There is a risk with prey animals becoming socialized with humans. If done improperly in a manner that startles or induces fear, the young animal may develop and retain a fear of being around humans.

FLIGHT ZONE

Members of all species have invisible borders, *flight zones*, around them in the presence of possible danger. The diameter of an animal's flight zone for humans varies by the animal species involved, its breed, the amount of prior exposure to humans, the quality of prior contact with humans, and the age of the animal when exposed to humans.

MAJOR REACTIONS TO INVASION OF FLIGHT ZONES

- **Escape:** Invading its flight zone will often cause an animal to attempt to escape.
- **Freezing:** Invading a flight zone does not always result in flight of the animal. Some may be willing to fight, and some may have tonic immobility, i.e., they freeze with fear. Freezing is more common in prey animals, like rabbits.

MINOR REACTIONS TO INVASION OF FLIGHT ZONES

Less severe dissociative behaviors that can be induced by fear are intense grooming in cats and repetitive yawning in dogs. Dogs may exhibit an *adrenaline shake off* from a relief of fear after an intruder exits the dog's flight zone.

Animal senses activate and modify their behavior.

OLFACTORY (SMELL)

The sense of smell is more acute in all domestic animals than in humans. Animals monitor the odor of urine, feces, sweat, breath, and special skin organs, such as anal glands in dogs, to identify others, assess their status in a reproductive cycle, and determine their social rank.

Cologne or other pungent cosmetic odors can cause animals to resist handling and restraint.

- **Dogs:** Dogs have the keenest sense of smell of any domestic animal. Dogs can detect odors that are 10,000 to 100,000 times fainter than what the human nose can detect.
  - **Communicating Odors:** Some communications among dogs are by emitted pheromones from their body by secretions of saliva, urine, feces, and anal sacs.
  - **Scent Detection Services:** Dogs can be trained to detect explosives, corpses, drugs, among other odoriferous objects by using their extraordinary ability to smell.
- **Cats:** Cats have scent glands under the chin, corners of the mouth, side of the forehead, and between their toes. They also emit odors through urine and feces.
  - **Marking Territory:** Urine spraying and odor from their front pads, which is left when scratching objects, are used to mark a cat's territory.
  - **Facial Marking:** The small cheek glands, near the corners of their mouths, are used to leave odors after rubbing (called *bunting*) on objects, including people, that they perceive as their territory (Figure 2.1).

Figure 2.1   Bunting to smear odor on a favored human.

- **Reptiles and Birds:**
    - **Reptiles and Food:** Reptiles become excited at the smell of food. If the smell is on a handler's hands, the odor can entice a reptile to bite a hand.
    - **Birds and Toxic Air:** The respiratory system of birds does not provide many of the protections of the respiratory system of mammals against airborne insults. Birds are particularly sensitive to odors, and some can be lethal to birds. Canaries have been used to monitor for harmful gases in mines.
- **Vomeronasal Organ**
    - **Location and Structure:** The vomeronasal, also called Jacobson's, organ is located in the roof of the mouth. It consists of two sacs that are connected to the nasal cavity by fine ducts. It is found in many mammals and all snakes and lizards (Figure 2.2).
    - **Purpose:** When domestic mammals smell sexual odors, many will lift their upper lip and open their mouth, a procedure called the *flehmen response*. The purpose of the flehmen response is to increase the opening of the ducts that carry the smell to the nasal cavity and the olfactory membrane. This enhances the detection of the odor.
    - **Manifestations:**
        - **Gape:** Cats *gape* (mouth open with tongue placed behind upper incisors) when smelling other cats' urine.
        - **Flicking Tongue:** Snakes smell using their forked tongue to collect particles in the air. The tongue then pulls the particles into the mouth, where they are dipped into the vomeronasal pits in the roof of the mouth.
- **Aromatherapy for Handling Animals**

In nature, odors have effects on animal behavior. Their ability to alter behavior to facilitate handling is less clear.

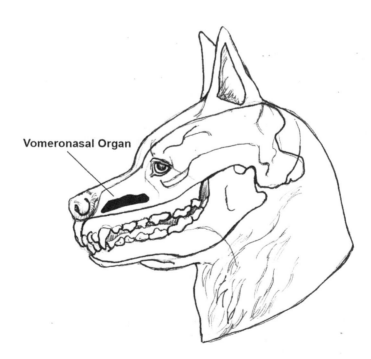

Figure 2.2   Vomeronasal organ in a dog.

- **Pheromones:** Pheromones are chemicals used for communication by smell.
  - **In Nature:** Natural pheromones are well-established important communicators of individual identity and reproductive status in many, if not all, species.
  - **Synthetics:** Synthetic pheromones and essential oils have been proposed to be effective means of calming dogs and cats.
  - **Efficacy:** Aromatherapy is a form of alternative therapy, similar to nutraceuticals, that is not required to prove efficacy to be marketed. Claims of efficacy are usually based on anecdotal statements or small studies without sufficient controls and independent evaluations that are required of pharmaceuticals.
  - **Short Duration Distraction:** Synthetic pheromones may be more of a means of nonthreatening distraction than a mind-altering drug. Those that appear to affect cat behavior have a short duration of effects, less than 30 minutes.
- **Aromatherapy Agents**
  - **Catnip:** Nepetalactone is a volatile oil from the catnip plant (*Nepeta cataria*), a member of the mint family. It is an attractant for about three out of four cats. The playful activity it evokes in cats temporarily causes distraction from other nonthreatening stimuli.
  - **Silver Vine and Other Volatile Oil Plants:** Silver Vine (*Actinidia polygama*), and to a less extent, Tatarian honeysuckle (*Lonicera tatarica*) and valerian root (*Valeriana officinalis*) contain volatile oils with similar effects to catnip on most cats.
  - **Feline Synthetic Facial Pheromone:** Facial glands in cats produce a pheromone involved in bunting (facial rubbing) to mark possession. A synthetic facial pheromone of cats in an alcohol solution is a popular aromatherapy intended to calm cats, although independent controlled studies on behavior and physiologic parameters have yet shown no effect.
  - **Dog Appeasing Pheromone:** Dog appeasing pheromone (DAP) is a pheromone produced by the skin of the mammary gland of dogs after giving birth and during the nursing period. DAP is believed to aid in bonding pups to the mother. Synthetic DAP is an aromatherapy purported to calm adult dogs but with mixed efficacy results.
  - **Lavender and Chamomile Oil:** Oils from the plants lavender and chamomile are proposed to have a calming effect on small animals.
- **Distracting Odors:** Odors from handling environments and the handler's body can be distracting, even threatening, to animals. Therefore, it is a good practice to avoid or mitigate odors when handling animals (Table 2.2).

HEARING

Sounds are an important communication method and stimuli that warn animals of potential danger.

- **Auditory Communication:** Animals are able to differentiate each member of their group's voices. Similarly, they know each of their handlers' voices. They are able to recognize and associate sounds that occur with feeding, distress, and breeding, among others.

Table 2.2 Avoiding Distracting Odors
• Use unscented hydrogen peroxide commercial environmental cleansers or one-half 3% hydrogen peroxide and one-half water mixture for sanitation.
• Avoid perfumed body products.
• Keep food and other animal odors off hands and clothes.

- **Three Aspects of Hearing:** There are three aspects to hearing sounds: *intensity, frequency,* and *directional ability.* Intensity (amplitude) is measured on a logarithmic scale in units called decibels. Frequency is the number of vibrations per second.
  - **Intensity:** Low-toned soft sounds are soothing to animals. Soothing background music may calm animals and is often used in veterinary hospitals and kennels at least to the benefit of animal owners.
    - **Music:** Music has been promoted to calm dogs. Disparate (soft rock, reggae, and classical) music has been recommended by different studies that did not use independent evaluators blinded to the presence of music. The benefits of music on calming dogs is disputable, and assumptions of benefits may be based on anthropomorphism.
    - **Level of Proof:** Based on nonblinded qualitative assessments of *handling scores* and *cat stress scores,* cats have been claimed to benefit from listening to *cat-specific music.* Quantitative assessment of physiologic parameters were unaffected by cats listening to music.
  - **Frequency:** All domestic mammalian animals can hear higher-frequency sounds than humans. Measurement units are Hertz (Hz).
    - **Dogs and Cats:** The auditory range of humans is approximately 20 to 20,000 Hz, while dogs have a range to around 45,000 Hz, and cats can hear up to 75,000 Hz.
    - **Rodents:** Rodents have a higher upper range of 75,000 to 80,000 Hz, similar to that of their chief predator, the domestic cat.
    - **Birds and Lizards:** Birds range of hearing is similar to humans, while lizards react best to lower-frequency sounds below 5,000 Hz.
    - **Snakes:** Snakes have internal ears that detect sounds only if the sound causes low-frequency vibrations. They feel the vibrations with their jaw from the surface that they are lying on, which are then transmitted to their internal ears.
  - **Direction:**
    - **Assessing Point of Attention:** The ability to hear can be enhanced by moving external ear position in the direction of the sound source. Ear position is an indication of expected behavior. For example, each external ear in dogs has 18 muscles to enable it to position for optimal hearing.
    - **External Ear Type:** Erect external ears are best at detecting sound direction and amplifying sound to the eardrum. Dog breeds with erect ears are the most often used as sentry animals to guard property. Some breed standards expect ears to be trimmed so that they can be made erect to assist hearing.

VISION

Vision is the primary sense used for detecting danger for many species (Table 2.3). Impaired vision can affect an animal's tractability. For example, cat muzzles block vision and calm cats, diurnal birds are more easily handled in subdued lighting, and raptors are more calm with hoods on.

Table 2.3 Vision Components
• Field of view
• Depth perception (stereopsis, i.e., judgment of distances)
• Acuity (focus)
• Perception of motion
• Color differentiation and night vision

- **Field of View:**
  - **Predators:** The field of view for predators is narrower than in prey animals.
    - **Dogs:** Dogs have a horizontal field of view of approximately 240 degrees, slightly wider than in humans.
    - **Cats:** Cats have a horizontal field of view similar to dogs (200 degrees) but a wider vertical field of view due to their vertical pupils.
  - **Prey:** The eyes of grazing/prey animals are located on the sides of their heads and protrude slightly in comparison to predator eyes. The side location and protrusion of their eyes allow prey animals even greater horizontal peripheral vision than that of predators. For example, horses have a 355-degree horizontal field of view.
- **Pupil Shape and Light Accommodation:**
  - **Pupil Shape:** Pupil shapes vary with species' vision requirements. Small animals have round or vertical pupils (Figure 2.3).

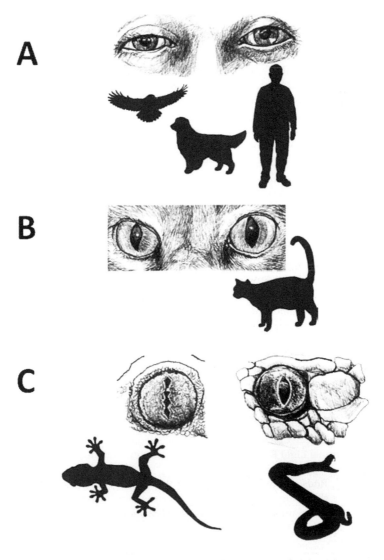

Figure 2.3 Shape of ocular pupils in animals.

- **Dogs:** Dog ancestors, lions, tigers, and birds of prey are predators that monitor their prey from a distance and hunt while moving. This type of hunting is facilitated by round pupils.
- **Domestic Cats, Venomous Snakes, and Lizards:** Small cats, like the domestic cat, venomous snakes, and lizards are low-profile hunters. These animals have vertical pupils, which may permit an enhanced ability to focus through thin vertical gaps in tall grass, where they frequently hunt. They have large corneas and pupils for their body size.
  - **Speed of Accommodation to Light:** Dilation and constriction of the pupil are the primary means of accommodating to changes in lighting. Bird pupils and pupils in crepuscular and nocturnal animals, such as cats, accommodate relatively rapidly.
- **Depth Perception:** Depth perception requires overlapping fields of vision from each eye.
  - **Humans and Predator Animals:** Humans and predator animals have binocular vision, focusing on objects of interest with both eyes. The central overlap that permits depth perception in dogs and cats is about one-half that of humans but better than prey animals. Cats binocular vision is 140 degrees.
  - **Prey Animals:** Prey animals have a wider total field of view than predators, but less is binocular. Horses' binocular vision is 65 degrees which limits their depth perception.
- **Acuity and Perception of Motion:**
  - **Visual Acuity:** Visual acuity is the ability to see details.
    - **Comparative Acuity:** Domestic mammals do not have the visual acuity of humans. Near vision is relatively poor. Normal humans have 20/20 acuity.
    - **Dogs and Cats:** Dogs are estimated to be 20/75 (normal humans can see clearly at 75 feet with the clarity that dogs see at 20 feet). Cats have 20/100 acuity.
    - **Birds:** Birds have exceedingly good visual acuity. Their lens is flexible, which aids their ability to rapidly focus on objects.
    - **Reptiles:** Visual acuity of reptiles is poor. However, reptile vision can vary with families depending on the species' lifestyle. For example, arboreal snakes have better vision than terrestrial snakes that burrow.
  - **Detection of Motion:**
    - **Predators:** In most predators, the area of greatest acuity is a circular area in the retina, called the *fovea* or *area centralis*. To visually evaluate the greatest detail, predators have to hold their head still and concentrate the image on the fovea.
    - **Prey:** In contrast, grazing prey animals, such as rabbits, have a *visual streak*, an elongated band that runs across the retina. This permits them to better detect motion in their peripheral vision.
- **Color Differentiation and Night Vision:** Most animals see better in low light than do humans but perceive fewer colors.
  - **Retinal Light Receptors:** The retina of the eye contains two types of light receptors: rods and cones.
    - **Rods:** All mammals have more rods than cones, and animals have more than humans. Because rods perceive lower-intensity light than cones, rods aid in night vision. Nocturnal mammals having a preponderance of rods may be unable to distinguish colors.

- **Cones:** Cones perceive objects best in bright light and can differentiate colors. The area centralis or visual streak contains the highest concentration of cones and the lowest concentration of rods.
- **Color Vision in Mammals:**
  - **Humans:** Humans have three types of cones, which permit trichromatic color vision (tones of red, green, and yellow).
  - **Domestic Animals:** Most domestic animals that are active during daylight have two types of retinal cones and dichromatic color vision (yellow and blue). They cannot distinguish colors in the range of 510 to 590 nm, the red wavelength.
  - **Dichromatic Color Vision:** Animals with dichromatic vision appear to see blue and yellow best and have trouble perceiving red and green. With dichromatic vision, red is dark and green is light gray. Dichromatic vision may aid in seeing sudden movements and objects in low light.
- **Color Vision in Reptiles and Birds:** Reptiles and tropical birds have four types of retinal cones and may perceive more colors, or colors in dim light, than humans can see. Birds can see ultraviolet, blue, green, and red (tetrachromatic vision). Bird vision peaks in the orange-red portion of the spectrum.

- **Night Vision:**
  - **Retinal Rod Cells:** Rod cells are responsible for magnifying light impulses. Rods are able to detect low-intensity light and motion and differentiate shades of gray, but they provide poor resolution. Most domestic animals, especially nocturnal predators (cats), have many more rods than do humans.
  - **Tapetum Lucidum:** Species that are scoptic (have vision in dim light) or nocturnal also have a tapetum lucidum (reflective structure in the retina that increases the gathering of light). This results in superior night vision and more intense differences in grays, plus better detection of motion. The tapetum increases light detection at night up to seven times in cats.

TACTILE (TOUCH)

Tactile sensations play an important role in communication among animals and individual spatial orientation.

- **Communication:** Animals communicate with each other with a range of touches. Soothing or grooming touches reinforce the bonding within a group. Moderately firm stroking conveys a better impression of confident leadership of a good handler.
- **Spatial Orientation—Vibrissae:**
  - **Definition:** Whiskers (vibrissae) are large, long, well-innervated hairs surrounded by a vascular sinus.
  - **Location:** Most species have vibrissae on the upper (maxillary) lips. Dogs also have supraorbital (above the eyes), genial (cheeks), and interramal (between the angles of the mandible). Cats also have carpal vibrissae.
  - **Function:** The function of vibrissae is to feel spatial limits, air movements, and movement of captured prey.

BODY HEAT

Some snakes (pit vipers, pythons, and some boa constrictors) have infrared receptors found in loreal pits for tracking warm-blooded mammals (Figure 2.4). The pits with heat receptors may be between the nostrils and eyes, as with pit vipers, or just below the nostrils, which is the case in pythons. The snake's brain is believed to form images from infrared rays in a way similar to visual images from the eyes.

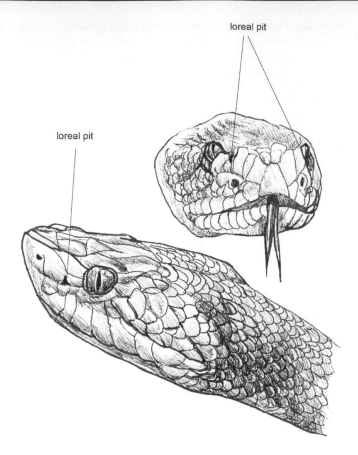

loreal pit

loreal pit

Figure 2.4 Loreal pits in some snakes with infrared receptors.

RESISTANCE BEHAVIORS

Resistance to handling may be manifested as avoidance or aggression. Aggressive behavior can have many possible causes (Table 2.4).

MATERNAL AGGRESSION
Animal mothers will protect young when they may not be willing to protect themselves. This is most acute after birth to the time of peak milk production. In dogs, this is the first 3 weeks after birth.

PAIN-RELATED AGGRESSION
Feeling pain or just anticipating pain can cause aggression.

- **Perception of Pain:** Pain-related aggression is a natural response to noxious stimuli. A major reason for people being bitten by dogs is trying to capture or move an injured dog that is in pain.
- **Anticipation of Pain:** Anticipation of possible pain can induce fear-aggression. Frequent gentle handling of all parts of an animal's body, with no other purpose than training, can desensitize animals to anticipated pain aggression.

Table 2.4   Causes of Aggression
•	Maternal
•	Pain
•	Predatory
•	Territorial (Possession)
•	Fear-induced
•	Intermate
•	Dominance

PREDATORY AGGRESSION

Retreating from an aggressive predator can be perceived as fleeing prey and aggravate aggression.

- **Dogs:** Examples of predatory aggression in dogs are children running from strange dogs and dogs chasing and biting joggers and bicyclists.
- **Cats:** Predatory aggression in cats is characterized by a low posture, crawling, and freezing in place. Playing with cats by inviting them to pounce on wiggling fingers or toes can stimulate predatory aggression.

TERRITORIAL AND POSSESSION AGGRESSION

Territorial and possession aggression are common causes of dog bites.

- **Territorial Aggression:**
  - **Dog and Cats:** Dogs or cats may establish their cage, run, yard, or family car as personal territory and become aggressive if they anticipate being separated from their territory. Some dog breeds are more prone to territorial aggression, such as Rottweilers and German shepherd dogs, which have been selectively bred for the trait of territorial aggression.
  - **Hamsters:** Hamsters are aggressive in protecting their territory. Males have scent glands on their backs near the hips that are used for territorial marking. Adult golden hamsters need to be housed separately to prevent fighting.
- **Toy Aggression:** Some dogs can become very possessive of their toys and react with aggression if threatened with separation from their possession.
- **Food Aggression:** Protection of food while eating is a common possession aggression among dogs and cause for children being bitten when near dog food bowls. Dogs should be trained to control that tendency while they are in their critical socialization period prior to 16 weeks of age by not feeding puppies in communal bowls and occasionally hand-feeding from the bowl.

FEAR-INDUCED AGGRESSION

Fear-induced aggression in dogs or cats is very common in veterinary hospitals.

- **Body Language:**
  - **Dogs:** Fear-based body language in dogs is a fixed stare, rigid neck, head lowered, ears laid back, lips pulled back, and tail between the legs.
  - **Cats:** Cats crouch with their ears back, wrap their tail close to their body, raise the hair on their back, and hiss combined with a low rumbling, throaty sound.
- **Risk of Intensifying Aggression:** When the fear-inducing stimulus (a person) backs off, the aggression is rewarded, and the animal will intensify its aggression the next time it feels threatened.

INTERMATE AGGRESSION

Female animals in estrus will intensify male agitation and aggression. Male mice, rabbits, or bearded dragon males should not be housed with other males of their species because of the risk of intermale aggression. Feral tomcats may kill young kittens from another male when taking over a new clowder (group of cats).

DOMINANCE AGGRESSION

Potential to dominate may occur from selective breeding for an animal with that trait and be intensified by the influence of male hormones and larger body size (Table 2.5). A dog that is aggressive in an attempt to establish dominance over a human is dangerous and must be handled initially with higher levels of restraint by experienced handlers.

- **Body Size:** A major factor of establishing dominance in a group is body size. For example, dogs attempting to dominate another will raise the hair on their back and elevate their stance in front to appear larger. Larger individuals are more dominant than small individuals of the same breed, and larger breeds are more dominant to smaller breeds.
- **Male Sex Hormones:** Male sex hormones are also major influences on displays of dominance. Castration, prior to puberty, does not eliminate the possibility of dominance aggression, but it significantly reduces it. Postpubertal castration effects are less impressive since adult male behavior may become ingrained on the nervous system before the removal of male hormone stimuli.
- **Selected Trait for Aggression:**
  - **Natural Genetic Selection:** Normally dominant animals only become aggressive if they have to protect themselves or their group. Within a species, animals do not risk serious injury or death to establish dominance, except for mature males during breeding seasons.
  - **Selective Breeding:**
    - **Dominance Aggressive Breeds:** Dominance aggression is more common in some dog breeds than others. For example, it is relatively common in spaniels, Belgian Malinois, German shepherd dogs, Doberman pinschers, Rottweilers, pit bulls, Akitas, and chow-chows.
    - **Early Signs of Dominance Aggression:** Signs of dominance aggression in dogs appear in early adulthood (1–3 years of age) and are manifested when guarding food or toys, being overprotective of some family members, or growling and snapping when told *no*.
- **Domination Techniques in Training:** Punishment or domination techniques, such as holding a dog on his back and staring at him (alpha-rolls), often make a fear-aggressive animal more aggressive.
- **Suppression of Dominance:**
  - **Require Earned Attention:** Dogs tend to seek human attention. Ignoring their attempts to get their handler's attention is the first step in establishing handler dominance, followed by offering attention on the handler's terms and praise only if obedient to the handler's commands.
  - **Immediate Response to Simple Commands:** Performing simple commands such as *sit* or *down* for dogs places them in submissive positions and reinforce a handler's status as being their leader which, in turn, increases his safety in handling them.

Table 2.5	Major Factors in Exhibiting Dominance Aggression
•	Selective breeding for dominance tendency
•	Larger-than-typical body size
•	Influence of male sex hormones

There are species differences in typical aggression and avoidance.

DOGS

Dogs elevate their position if aggressive and lower their body if submissive.

- **Dominance Body Language:** Dogs' dominance posture is head up, tail up, front end is held high, rear end may be slightly crouched, teeth may be bared, and stare is fixed (Figure 2.5).
- **Avoidance Body Language:** Submissive posture is front end lowered, eyes do not make direct contact with handler's eyes, tail is between legs, submissive urination or defecation may occur, and the dog may roll on its back to expose abdomen (Figure 2.6).

Figure 2.5 Aggressive posture in a dog.

Figure 2.6 Submissive posture in a dog.

- **Triggering Environment:** Some dogs are aggressive in the presence of their owner; others, especially dogs strongly bonded to one person, may be more aggressive away from their owner. This is not true of most dogs, however.

CATS
Cats have uncanny speed over short distances that allows them to stalk prey and then move for a rapid kill before they can move.

- **Anxiety Sign:** A common sign of irritability or anxiousness is flicking the end of the tail.
- **Avoidance First:** Most cats will attempt to escape rather than fight with a handler. They are escape artists, so all exits from a handling room must be closed before handling cats. They can be highly aggressive if they feel endangered and trapped.

OTHER SMALL MAMMALS
Most aggression and efforts to dominate are within a small mammal's own species and related to breeding behaviors. Other than ferrets, most of their life is directed toward avoidance of possible predators.

COMPANION BIRDS
Companion birds seek higher perches to establish dominance. Handlers should not permit companion birds to sit on their head or shoulders, since this encourages dominance behavior toward the handler.

REPTILES
Innate aggressiveness varies widely among reptile species. The presence of food or the process of shedding can intensify aggressiveness in many reptiles.

- **Species:** Larger species of lizards and snakes are generally more aggressive than small species. Lizards indigenous to arid lands are more docile than iguanas, which come from tropical forests.
- **Food:** Lizards and snakes become aggressive in the presence of food or odor of food. Handlers should not handle food prior to handling reptiles. Feeding cages or boxes should be used to feed lizards and snakes rather than their primary cage to reduce aggression in their primary cage.
- **Shedding:** All reptiles regularly shed their skin by a process called *ecdysis*. Snakes typically shed their skin in its entirety at one shedding, which can impair their sight and make them less tolerant of handling.

TRAINING METHODS FOR HANDLING ANIMALS

Training animals to be handled should be the major part of preparing young animals to become socialized with humans. The most effective means of training depends on species and what is to be learned. Success at training is dependent on animal genetics, observational learning from its mother, and the quality of handling it receives prior to puberty (Table 2.6).

Table 2.6	The Foundation for Success at Handling and Training Animals
•	Selection of a young animal that is an offspring of parents that have been willingly receptive to their own training
•	A mother that has been properly socialized to humans and is quietly handled in the offspring's presence (observational training of offspring)
•	Gentle repeated handling of the young animal during its critical socialization period

REINFORCEMENTS

Either *positive reinforcement* (adding a reward such as food treats) or *negative reinforcement* (removal of a noxious stimulus) are used to instill trained behaviors. Food treats are used as positive reinforcement for desired behavior, and sharp, brief reprimands, vocal or physical, such as a tug on a training collar when needed are negative reinforcement to correct inattention or misbehavior.

- **Timing Is Critical:** The timing of either reinforcement is critical, i.e., within three seconds of the behavior to be reinforced with a treat or discouraged with a reprimand. Late negative reinforcement becomes simply punishment, which can lead to the animal becoming aggressively defensive or developing a helplessness attitude that inhibits further learning.
- **Positive Reinforcement:**
  - **Dogs and Cats:** Small predatory companion animals (dogs and cats) can be trained effectively with positive reinforcement. Their natural behavior is to investigate (hunt) sources of food and be rewarded by food when the hunt is successful.
  - **Reinforcement Treats:** Treats for positive reinforcement of dogs should be the size of a pea to a grape and able to be eaten quickly. It should be a treat that the dog does not get at any other time. Food treats should be gradually supplanted with tactile and voice rewards, particularly if a dog is to become a working dog, guardian, or support companion. Treats should be offered at the level of the muzzle or lower to avoid unintentionally encouraging the dog to jump on the handler.
- **Negative Reinforcement:**
  - **In Nature:** Misbehavior within small predators is corrected by immediate warning (growl or hiss), followed, if needed, by a sharp, brief vocal or physical reprimand by a more socially dominant member of the group.
  - **In Training:**
    - **Pro:** Negative reinforcement, such as verbal reprimands or tugs on leashes, can be preferable in some instances of handling dogs or cats due to more rapid timing of delivery than positive reinforcement and less ability for the dog or cat to ignore the stimulus.
    - **Con:** Negative reinforcement has the potential of being aversive, so it should never be harsh enough to lead to avoidance behaviors, such as the dog being unwilling to engage the handler visually. Reprimands must be in frequent, metered to the situation, and not injurious.

SHAPING AND CHAINING

Advanced training of maturing or mature domestic animals involves *shaping*, breaking a task to be learned into small pieces, which are gradually refined. The refined small pieces of a trained activity are then performed in sequence, a process referred to as *chaining*.

COUNTERCONDITIONING

Counterconditioning is rewarding no response and is useful in training animals to accept something that might cause a fearful reaction. Counterconditioning is beneficial in training dogs and cats to accept veterinary hospitals, veterinary examinations, nail trims, blood collection, injections, and transport crates. Rewards are typically highly desired food treats, such as chicken or turkey baby food, peanut butter, braunschweiger, or squeeze cheese.

HABITUATION AND DESENSITIZATION

Habituation and *desensitization* are methods of reinforcing no response to certain stimuli. Habituation is providing a steady stimulus that causes an undesired response until no response occurs. Desensitization is using a repeated stimulus with increasing intensity until no response occurs.

AVERSIVE TRAINING METHODS

Aversive training methods should be avoided, but sometimes what is aversive is conditional. For example, choke (training/attention) or prong collars are used in dogs to regain a distracted dog's attention. Choke collars and prong collars can be aversive, but they are not if used with the correct timing and with the minimum effort to elicit an intended response.

HEALTH AND BEHAVIOR

Illness or injury can markedly alter the animal's tolerance of handling.

ASSESSMENT OF HEALTH

- **Effects of Illness on Behavior:** Illness may cause *anhedonia*, a reduction of normally pleasurable behaviors, such as self-grooming, allogrooming, playing, or exploring.
- **Health Assessments:** Before handling an animal, a handler should always attempt to observe practical assessments of the animal's health (Table 2.7).

PREDATOR VERSUS PREY HEALTH BEHAVIORS

- **Predators:** It is natural for predator animals to become defensively aggressive if ill or in pain. Common maladies are injury, arthritis, skin sores, and febrile infections.
- **Prey:** Prey animals tend to attempt to hide their illnesses. Some will fake eating and minimize lameness. They will become more social so they do not stick out from the group.

STEREOTYPIC BEHAVIOR AND ENVIRONMENTAL ENRICHMENT

The determination of psychological stress in animals is qualitative. This tempts some people to anthropomorphize that certain situations are mentally stressful to animals, which may or may not be the case.

IMPERFECT METHODS TO QUANTIFY STRESS

Efforts at quantitative measures have traditionally relied on the measure of cortisol levels and heart rates, both measures of parameters that change within minutes and are affected by multiple stimuli.

PRAGMATIC MEANS TO MEASURE STRESS

- **Visible Changes to Normal Behavior:** Visible reactions of animals to situations that induce apprehension, fear, trust, respect, and pain are very similar to observable human reactions. Experienced animal handlers can recognize these primitive basic feelings in

Table 2.7 Practical Assessments of Animal Health
• The animal's overall appearance (body condition, hair/coat condition)
• Locomotion
• Interactivity with other animals
• Consciousness of the environment
• Evidence of food and water consumption
• Presence of fecal matter and appropriateness of the character of feces relative to the animal's own species
• Appearance and quantity of voided urine.

animals as well as any human can recognize the visible signs of fear, pain, and other basic reactions in another human.

- **Recognition of Stereotypic Behaviors:**
  - **Ethograms:** An ethogram is a catalog of behaviors exhibited by an animal. Changes in grooming, foraging, resting and sleeping, attempts to thermoregulate in an environment, or playing can be used to assess welfare in animals.
  - **Dysfunctional Behaviors:** Failure to appropriately thrive and repetitive alterations in normal behavior are reliable indicators of stress in animals. *Stereotypic behavior* is dysfunctional behavior, usually induced by stress and influenced by genetics.
  - **Causes of Stress:** The stresses are often excessive confinement; barren, boring environments; or isolation from their own species. Animals raised in a barren environment during their socialization period are more prone to stereotypic behaviors.
  - **Manifestations:** Stereotypic behaviors are characterized by repetitive actions having no obvious purpose. These behaviors are not seen in wild animals or those with relative freedom and adequate stimuli for mental exercises. Common types are usually forms of pacing or oral behaviors (Table 2.8).
  - **Prognosis:** Observed stereotypic behaviors may relate to current environment or past environment. Once the behavior has been established, it may be permanent due to alterations in primitive brain locations.

NUTRACEUTICALS FOR CONTROL OF ANXIETY

EFFICACY AND SAFETY OF NUTRACEUTICALS

Nutraceuticals are over-the-counter food substances administered orally with the intention to improve health or have medicinal benefits but generally without objective proof of efficacy. They are not considered pharmaceuticals and therefore are not regulated the same as drugs. Nutraceuticals are regulated by the U.S. Food and Drug Administration's reaction to public complaints involving egregious medical claims, misleading marketing, and contamination with hazardous materials.

ANTIANXIETY NUTRACEUTICALS

- **L-Tryptophan:** Tryptophan is a large amino acid that is a precursor to serotonin. Tryptophan dietary supplements are administered with the expectation that serotonin levels in the brain will increase, improving mood and behavior. However, the clinical benefits in humans are insignificant, and behavior improvement in treated dogs has not been scientifically validated.
- **Melatonin:** Melatonin is a hormone produced by the pineal gland in the brain and also by plants. Oral melatonin administration is proposed to treat insomnia in people and modulate fear in dogs, but proof of efficacy is currently insufficient.
- **Alpha-Casozepine**: Alpha-casozepine is a trypsin hydrolysate of the mammalian milk protein casein. It has been associated with a decrease in signs of anxiety in dogs and cats, but the evidence is weak, and more studies on clinical efficacy are needed.

Table 2.8	Example Stereotypic Behaviors
•	Spinning and wall bouncing in dogs
•	Circling and pacing in mice
•	Feather pulling in caged birds

- **L-Theanine**: Theanine is an amino acid found in tea leaves. Reduction of anxiety in dogs and cats has been attributed to the oral administration of theanine, but more studies are needed to confirm these claims. Because health claims in humans were not substantiated by the European Food Safety Authority, the European Union now prohibits L-theanine health claims for humans.
- *Souroubea* and *Platanus* : A mixture of *Souroubea* spp vine and *Platanus* spp bark has been promoted as a source of betulinic acid, purported to have anxiolytic effects. Independent controlled efficacy studies have not been reported.
- *Bifidobacterium longum* : *Bifidobacterium longum* is considered a psychobiotic, which is a live bacteria that is theorized to have effects on behavior via the gut–brain axis. Independent controlled efficacy studies have not been reported.
- **Passionflower:** A dried alcoholic extract of the climbing vine passionflower can cause sleepiness and ataxia. Its ability to reduce anxiety has not been proven scientifically. Its safety in small animals is not known.
- **Chamomile**: This herb has been proposed to suppress anxiety. Its safety in small animals is unknown.
- **Valerian:** Valerian is also an herb promoted for the control of anxiety in humans based on anecdotal evidence. Scientific evidence is inconclusive. Its safety in small animals is unknown.
- **Cannabidiol:** Cannabidiol (CBD) is one of more than 100 cannabinoids in the *Cannabis sativa* (marijuana) plant. If purified, it does not have the psychoactivity of tetrahydrocannabinol (THC).
  - **Proposed Benefits and Known Risks:** CBD is marketed as a calming and pain-relieving agent, but independent controlled efficacy studies are lacking. However, depression, liver damage, diarrhea, and altered metabolism of some concurrently administered medications are established potential adverse effects.
  - **Confusing Regulations:** Because the regulation of cannabis products in the U.S. is a fluid, conflicting, and confusing mess, false claims of medicinal effects and mislabeled products containing little to no CBH are common.
  - **Federal versus State Regulations:** Cannabidiol extracted from marijuana is classified federally as a Schedule I Controlled Substance. It is not approved as a prescription drug or a dietary supplement. However, some states have passed laws to allow CBD sales, while federal enforcement of sale as a federally controlled drug is inexplicably ignored.
  - **Recommendations Allowed:** Veterinarians may recommend CBD veterinary products under the Animal Medicinal Drug Use Clarification Act.

ANIMAL BEHAVIOR SPECIALISTS

Knowing basic animal behavior is essential for anyone to become a good animal handler. Animal behaviorists and animal handlers are not synonymous, however. Some excellent animal handlers may only be able to describe normal animal behavior in colloquial language, while some excellent animal behaviorists may be less-than-average animal handlers.

ANIMAL BEHAVIORIST TRAINING
There are no state or federal regulations for people to claim to be an animal behaviorist or trainer. Certification of training can be reputable, or not.

CERTIFICATION AS AN ANIMAL BEHAVIORIST
- **Veterinarians:** A certified animal behaviorist who is a veterinarian is the best source of information on how to diagnose abnormal behavior and what corrective measures to prescribe. The American College of Veterinary Behaviorists (ACVB) establishes the

requirements for formal education and evidence of acquired knowledge and skills for becoming a certified animal behaviorist.

- **Veterinary Technicians:** Certified or Registered Veterinary Technicians can go through specialty training to become a Veterinary Technician Specialist in Animal Behavior. The Academy of Veterinary Behavior Technicians (AVBT) certifies veterinary technician specialists in animal behavior.
- **Reputable Animal Behaviorist Organizations:** For more information on becoming a behaviorist or on behavior problems in animals, contact the ACVB www.dacvb.org or AVBT at www.avbt.net.

ANIMAL BEHAVIOR REFERENCES AND SUGGESTED READING

1. Bol S, Caspers J, Buckingham L, et al. Responsiveness of cats (*Felidae*) to silver vine (*Actinidia polygama*), Tatarian honeysuckle (*Lonicera tatarica*), valerian (*Valeriana officinalis*) and catnip (*Nepeta cataria*). BMC Vet Res 2017:13:70. doi:10.1186/s12917-017-0987-6.

2. Bowman A, Scottish SPCA, Dowell FJ, et al. The effect of different genres of music on the stress levels of kenneled dogs. Physiol Behav 2017;171:207–215.

3. Cobb ML, Iskandarani K, Chinchilli VM, et al. A systematic review and meta-analysis of salivary cortisol measurement in domestic canines. Dom Anim Endocrinol 2016;57:31–42.

4. Contreras ET, Hodgkins E, Tynes V, et al. Effect of a pheromone on stress-associated reactivation of feline herpesvirus-1 in experimentally inoculated kittens. J Vet Intern Med 2018;32:406–417.

5. Engler WJ, Bain M. Effect of different types of classical music played at a veterinary hospital on dog behavior and owner satisfaction. J Am Vet Med Assoc 2017;251:195–200.

6. Frank D, Beauchamp G, Palestrini. Systematic review of the use of pheromones for treatment of undesirable behavior in cats and dogs. J Am Vet Med Assoc 2010;236:1308–1316.

7. Hammerle M, Horst C, Levine E, et al. AAHA canine and feline behavior management guidelines. J Am Anim Hosp Assoc 2015;51:205–221.

8. Hampton A, Ford A, Cox RE, et al. Effects of music on behavior and physiologic stress response of domestic cats in a veterinary clinic. J Feline Med Surg 2020;22:122–128.

9. Hannah HW. Malpractice implications of animal restraint. J Am Vet Med Assoc 1999;214:41.

10. Herron HE, Shreyer T. The pet-friendly veterinary practice: A guide for practitioners. Vet Clin Sm Anim 2014;44:451–481.

11. Hewson C. Evidence-based approaches to reducing in-patient stress—Part 1: Why animals' sensory capacities make hospitalization stressful to them. Vet Nursing J 2014;29:130–132.

12. Hewson C. Evidence-based approaches to reducing in-patient stress—Part 2: Synthetic pheromone preparations. Vet Nursing J 2014;29:204–206.

13. Hewson C. Evidence-based approaches to reducing in-patient stress—Part 3: How to reduce in-patient stress. Vet Nursing J 2014;29:234–236.

14. Houpt KA. Domestic Animal Behavior for Veterinarians and Animal Scientists, 5th ed. Blackwell Publishing, Ames, IA, 2010.

15. Kim YM, Lee JK, Abd el-aty AM, et al. Efficacy of dog-appeasing pheromone (DAP) for ameliorating separation-related behavioral signs in hospitalized dogs. Can Vet J 2010;1:380–384.

16. Kipperman BS. The role of the veterinary profession in promoting animal welfare. J Am Vet Med Assoc 2015;246:502–504.

17. Kronen PW, Ludders JW, Erb HN, et al. A synthetic fraction of feline facial pheromones calm but does not reduce struggling in cats before venous catheterization. Vet Anaesth Analg 2006;33:258–265.

18. Masic A, Liu R, Simkus K, et al. Safety evaluation of a new anxiolytic product containing botanicals *Souroubea* spp. and *Platanus* spp. in dogs. Can J Vet Res 2019;82:3–11.

19. McConti L, Champion T, Guberman U, et al. Evaluation of environment and a feline facial pheromone analogue on physiologic and behavioral measures in cats. J Feline Med Surg 2017;19:165–170.

20. Mills DS, Ramos D, Estelles MG, et al. A triple blind placebo-controlled investigation into the assessment of the effect of dog appeasing pheromone (DAP) on anxiety related behavior of problem dogs in the veterinary clinic. Appl Anim Behav Sci 2006;98:114–126.

21. Ofri R. Vision in dogs and cats. Am Vet 2018;Aug:26–27.

22. Patronek GJ, Lacroix CA. Developing an ethic for the handling, restraint, and discipline of companion animals in veterinary practice. J Am Vet Med Assoc 2001;218:514–517.

23. Rushen J, Taylor AA, de Passille AM. Domestic animals' fear of humans and its effect on their welfare. Appl Anim Behav Sci December 1999;65:285–303.

24. Santos NR, Beck A, Blondel T, et al. Influence of dog-appeasing pheromone on canine maternal behaviour during the peripartum and neonatal periods. Vet Rec 2020;186:449.

25. Siracusa C, Manteca X, Cuenca R, et al. Effect of a synthetic appeasing pheromone on behavioral, neuroendocrine, immune and acute-phase perioperative stress responses in dogs. J Am Vet Med Assoc 2010;237:673–681.

26. Wells DL. Aromatherapy for travel-induced excitement in dogs. J Am Vet Med Assoc 2006;6:964–967.

3

DOGS

DOI: 10.1201/9781003110927-3

Dogs have been domesticated longer than any other species. They are in more U.S. households than any other animal. The proper name for male dogs is *dogs*. Females are *bitches*, and young dogs are *puppies*.

NATURAL BEHAVIOR OF DOGS

The great majority of domestic dogs are highly social. In their social interactions, dogs determine a social strata of dominance and submission. Dogs are generally protective of their territory, which radiates from a home.

BODY LANGUAGE

Communication among dogs involves body language, olfaction (feces, urine, anal glands, and other glands), and vocalization.

- **Normal Friendly Behavior** (Table 3.1):
- **Dominating Behavior**
  - **Initial Posture:** Dogs trying to communicate assertiveness or dominance aggressiveness try to appear as large as possible by piloerection (elevating the hair on the back and rump), standing with stiff elevated shoulders, lowering its pelvis with hind legs extended backward, and holding the tail high. The ears are forward, and the dog has a direct stare.
  - **Approach:** A dog's manner of approaching another dog or a human presumed to be less dominant is direct with its head and tail up. A submissive approach is to approach the other animal's side.
  - **Interaction:**
    - **Positioning Above:** To establish social dominance over another dog, a dog will jockey for a position above another dog by putting its head or forelegs on the neck or shoulder of a lower-ranking dog. Because of this, a handler lowering a hand onto a dog's head is perceived as attempting to dominate the dog.
    - **Growl and Smell:** A dominant dog may also circle and sniff with growls if the other dog moves. Olfactory communication is important among dogs. Nonfearful dogs approach new dogs and will immediately attempt to sniff the other dog's anogenital region.
  - **Territorial Marking:** Territorial marking, such as urinating on objects or where other dogs have urinated, is a means to establish dominance.
  - **Dominance Aggression:**
    - **Body Language:** An aggressive dog will stare at their opponent with lowered upper eyelids. Its lips will be drawn back and the mouth is held open. Ears are pointed forward, and the tail may slowly wag while elevated.
    - **Bold Actions:** Unlike dogs with fear aggression, dominance aggressive dogs will not hesitate to bite at a handler's face. Dominance aggressiveness is characterized by calculated actions, while fear aggression is reactionary.

Table 3.1	Normal Friendly Behavior of Dogs
•	Relaxed posture
•	Wagging tail and hips
•	Approaches with head down
•	Circles handler
•	May jump on handler's legs with front feet
•	May lie down and roll on back

- **Submissive Behavior:**
  - **Initial Posture:** Submissive dogs demonstrate a lack of direct eye contact with their ears back and their tail held low. A submissive dog may freeze in place or roll on its side and raise a hind leg to expose its belly.
  - **Approach:** If a dog approaches in a submissive manner, its body is curved toward the other animal or a person and the tail wags.
  - **Interaction:** Muzzle or face licking another dog is a submissive gesture. Some may lick the air as if face licking. Profound submission may lead to a submissive pose along with urination.
- **Fearful Behavior:**
  - **Initial Posture:** The ears are held back and as flat as possible next to the head. The neck is held rigid and the tail is tucked down and between its legs. Its head will be held down, and it may glance upward with its eyes at whatever it considers a threat. The dog may shiver or shake and lean away from the threat and snarl with its teeth exposed.
  - **Mannerisms:** Fearfulness is conveyed by repeated lip licking or yawning. Attempts to hide will occur when possible. If hiding is not possible, cowering in a corner is common, with the head held at shoulder level or lower. Freezing in place is common just before the dog attempts to bite at the threat.
- **Play Behavior:**

Dogs intending play will begin their interaction with another dog by assuming a *play bow* posture, i.e., rocking back on hind legs while lowering their front end by stretching forward with their front legs (Figure 3.1). The dog's ears are placed forward, and the tail is wagged rapidly. This posture is usually accompanied by a series of sharp barks.

Figure 3.1 Play bow posture.

VOCAL COMMUNICATION
Body language may be emphasized by vocal communications.

- **Growl:** Growling and snarling are intended to intimidate opponents.
- **Bark:** Barking is a sign of territory possessiveness or simply attention getting.
- **Whine:** Whining is a request for caregiving or affection. Whining may be accompanied by raising one front paw or pawing the animal or person of attention.

NATURAL BEHAVIOR BY BREED
The natural behavior of dogs has been modified genetically by selective breeding. These traits can be intensified or suppressed by training, but the trait will remain and can be manifested again under new circumstances such as a new home, owner, or handler, among many possibilities.

- **Personal Guard Dogs:** Personal guard dogs, such as the boxer, St. Bernard, and mastiff, tend to be even tempered and have a strong bond to family.
- **Livestock Guard Dogs:** Livestock guard dogs (e.g., Great Pyrenees, Komondor, Kuvasz) are solitary, bond less with handlers, and have low reactivity.
- **Herding Dogs:** Herding dogs (collie and shepherd breeds) bond strongly to individual handlers, have high desire to chase and herd things that move, and a low level of fear.
- **Terriers and Pinschers:** Terriers and pinschers are highly alert, aggressive, and develop possessive bonding with individual handlers.
- **Sighthounds:** Sighthounds (e.g., borzoi, greyhound, saluki, whippet) are aloof and quiet, have low reactivity, and bond less strongly with handlers.
- **Scenthounds:** Scenthounds (e.g., bloodhound, coonhound, basset hound, beagle) have low reactivity and low aggression with stoic dispositions.
- **Sled Dogs:** Sled dogs (e.g., malamutes, spitz, Norwegian elkhound, Siberian husky) are usually not aggressive but can be, bond weakly with owners, and have moderate reactivity.

SAFETY FIRST

HANDLER SAFETY
All dogs need to be exposed early in life to what their world will be like as an adult. Breeds of dogs that were selectively bred to guard property or livestock or to herd livestock were selected for an extra degree of assertiveness. It is the owner's responsibility to socialize and control dogs, particularly those with aggressive tendencies.

- **Socialization: The Key to Handling Ability:** Dogs and other domestic species go through an early socialization period, during which social experiences have a greater effect on the development of their temperament and behavior than if the experiences occur in later life.
  - **Timing:**
    - **Prime Socialization Period for Dogs:** In dogs, this period ranges between the end of the neonatal period, at 2 to 3 weeks (age when eyes and ears have first opened) to sometime between 12 and 14 weeks. Dogs that have little or bad experience with humans prior to 14 weeks of age rarely bond or respond to humans well for the rest of their lives.
    - **Social Maturity:** Social maturity occurs between 12 and 36 months of age, when adult behavior is best ingrained.
    - **Removal of Pups from a Litter:** Many social and behavioral deficits observed in adult dogs may be caused by removing puppies too early from the dam and littermates. Puppies need to learn social ranking between 3 and 8 weeks of age through play fighting, and how to interact with humans and other species from 5 to 12 weeks.

- **Preparation of Puppies for Socialization**
  - **Vaccinations First:** The American Veterinary Society of Animal Behavior recommends beginning socialization at 7 to 8 weeks of age and 7 days after first vaccinations and deworming treatment.
  - **Locations:** Classes should be held on surfaces that are easily cleaned and disinfected. Puppy exposure to dog parks, pet stores, or other areas that are highly trafficked by ill dogs or dogs of unknown vaccination status or are not sanitized regularly should be avoided.
  - **Shelter Dogs and Possibly Ill Dogs:** If adopted from a shelter, the puppy should be kept in their new home for 2 weeks before socializing with other dogs to reduce the risk of the puppy exposing other dogs to shelter-acquired diseases. Puppies should not socialize with other dogs that are sneezing, coughing, vomiting, or having diarrhea in order to reduce the risk of transmission of disease to the puppy.
  - **Diversity of Socialization:** Socialization should minimally include other people, children, other dogs, cats, vacuum cleaners, moving cars, bicycles, veterinary hospitals, and grooming parlors. A popular Rule of Seven is often applied (Table 3.2).
- **Puppy Classes**
  - **Groupings:** Ideally, puppies should be grouped by similar size. To limit distractions, a group should be no more than six puppies, and each puppy should have only one or two people handling it.
  - **Avoid Pain and Fear:** Puppies should never be exposed to an experience that is perceived as harmful, painful, or excessively frightening. If the puppy becomes apprehensive, its handler should give it a command and then reward it but not pet or cuddle the puppy immediately after it acts apprehensive, or it will interpret that fearful actions yield rewards.
  - **Activities Frequency:** Supervised play time should be scheduled each day. The play and training sessions should be short, about 15 minutes, and only 1% improvement expected each training session.
  - **Basic Commands:** All dogs should be taught basic commands in their first year of life for daily management, safety, and control of anxiety (Table 3.3).

Table 3.2	Rule of Sevens: by 7 Weeks of Age, Pups Should Have:
•	Been on seven types of surfaces
•	Played with seven different objects
•	Been in seven locations
•	Met with seven new and different people (young, old, disabled, different races, etc.)
•	Been exposed to seven challenges (similar to an obstacle course)
•	Eaten from seven different types of containers
•	Eaten in seven different locations

Table 3.3	Basic Commands Dogs Should be Taught in First Year of Life
•	Sit
•	Down
•	Stay
•	Come
•	Heel
•	Drop it
•	Go rest

- **Drag Line Leash:** When allowed play with freedom, a puppy's distracted attention can be regained as needed by having it wear a drag line leash at least 4 feet long.
- **Punishment:** Punitive methods, including scruff shakes, alpha rollovers, pinning to the floor, thumping the nose, swatting with rolls of paper, or shock collars should never be used for training puppies. Never call a dog to be reprimanded, or it will never learn or will unlearn the command *come*.
- **Leadership from Handlers:** Socialization with humans must present the handler as a consistent, gentle leader. Effective socialization must be one-on-one with each puppy, not as a litter. When attention is shown to the dog, the dog should reciprocate with its attention.
- **Role of Treats:** Positive reinforcements are initially small bits of dry food treats, which are combined with petting and other praises. Treats should also be small enough to be consumed in a couple of seconds.
- **Purposeful Interaction:** Petting should be reserved as only a reward for good behavior. Withdrawal of handler attention should be the penalty for poor behavior. Fearful behavior should not be rewarded with extra attention to try to comfort the dog, and apprehension should not be reprimanded.
- **Management of Anxiety:** The handler should have the puppy obey a familiar basic command such as *sit* and then reward it for sitting. A familiar situation, direct attention from a handler, and reward for appropriate behavior will provide distraction from its apprehension and promote a feeling of security for the puppy.
- **Social Rank of Handlers:** It is important for a handler to establish a superior social rank to the puppy's during its socialization. This requires controlling the puppy's resources and movements (*Procedural Steps 3.1*).

Procedural Steps 3.1	Establishing Proper Role of a Handler
1.	**Control Resources:**
	The handler should make the puppy sit before feeding.
	Place a hand in a food dish while it eats.
	Eventually feed it a portion of its meal by hand out of a feed bowl.
2.	**Control Movement:**
	Puppy movements should be controlled, as in being taught to sit if approached by strange people or when a stranger comes to the door.
	The puppy should be taught to wait for permission to go through doors or up and down stairs when on a leash.
	The handler should remain still, avoiding any attention to the puppy until its attention is directed only to the handler.

- **Response to Puppy Disrespect:** Puppies should not leave littermates and their mother until 8 weeks of age so they can better learn bite inhibition, among other beneficial social skills, from each other. A dog shows disrespect for a handler by putting its mouth around a hand or arm. If a puppy mouths a handler's arms, hands, or fingers, the handler should make a high-pitched sound and ignore the pup for about a minute before returning to more interactions with it.
- **Exposure to a Diverse World:** A puppy should experience a wide variety of people, animals, and situations in nonthreatening ways during their prime socialization period. Handlers should countercondition the puppy to being brushed, bathed, inspected, and having nails clipped and teeth and ears cleaned. This is accomplished by gentle, frequent, short-term handling sessions with small food treats whenever the puppy does not react adversely to the distracting stimuli.

Table 3.4	Basic Rules to Teach Children Under 8 Years Old in Handling Puppies
•	Only handle puppies if supervised by an adult.
•	Move slowly and be quiet and gentle.
•	Don't pick up puppies or try to carry them: sit and hold a puppy in your lap.
•	Never bother puppies when they are eating or resting in their crates.

- **Instruction of Children:** Handlers must always supervise interactions of puppies with other people, particularly children. Interactions need to be calm, gentle, brief, and controlled. Small children, in particular, should be closely supervised to insure against unpleasant or threatening experiences for the puppy (Table 3.4).
- **Exposure to Other Animals**
  - **Good Dog Models:** Exposure to other dogs during the socialization period should only be to good canine role models. The introduced dogs should at first be of similar size, friendly, healthy, and vaccinated dogs and other puppies, while larger and smaller dogs should be introduced later. Much is learned by puppies from observing how other dogs relate to humans and other animals.
  - **Cat Mentors:** Cats that are not aggressive nor afraid of dogs should be introduced to the puppy.
  - **First-Stage Location:** The first socializing with other animals, should be to other dogs that are introduced to a puppy's environment.
  - **Second-Stage Location:** The second stage of socialization is to take the puppy outside of its own environment to the homes of other friendly, well-behaved pets. A puppy should be socialized to any type of animal that it may come in contact with during the rest of its life, which, in some cases, may include birds, horses, cattle, sheep, swine, and others.
- **Commercially Bred (High-Volume) Puppies**
  - **No Required Socialization:** Commercially bred puppies are generally at high risk of inadequate socialization. Many states do not mandate socialization for commercially bred puppies, and those that do have vague requirements. Socialization is not a requirement for interstate shipment, and interstate shipment is permissible as early as 7 weeks of age.
  - **Outcomes:** Dogs obtained from pet stores are rarely socialized properly. They have significantly more aggressiveness toward humans, including family members, and other animals; separation-related problems, and inappropriate urination and defecation problems.
  - **Declining Industry:** Commercial-bred, high-volume puppy operations began with encouragement of the USDA after World War II as a supplemental income for farmers. Most states have never had sufficient regulations or enforcement of regulations to ensure the humane treatment of high-volume pet producers. Now, California and Britain have banned the third-party sales of commercially bred puppies, kittens, and rabbits, which is a trend that is expected to continue.
- **Potential for Injury**
  - **Overall Incidence**
    - **Injuries and Death:** Each year, dogs kill about 20 to 30 people in the U.S., seriously injure at least 800,000 with bite wounds serious enough to require hospital attention and are estimated to inflict a total of 4.5 million bites. Horses kill more people, but dogs cause more hospitalizations.
    - **Size of Dogs:** The most dangerous dogs are larger dogs, not because they necessarily bite more often but because their bites inflict more damage.

Most dogs involved in a killing are in the 50- to 100-lb. weight range. The ability to inflict a killing bite is instinctive.

- **Injuries:** Nonfatal wounds are usually to the arms, hands, or face. In addition to the puncture wounds, a large dog can generate enough pressure to cause significant crushing injuries. Tearing injuries may also occur when, after making the bite and holding on, the dog often shakes its head and sometimes its whole body or the victim tries to withdraw quickly.
- **Male Dogs:** Dogs involved in serious bites to humans are primarily male. Male dogs are 6.2 times more likely to bite people, and intact males are 2.6 times more likely to bite than neutered male dogs. However, once a dog develops the courage and ability to successfully bite a human, neutering has little effect in preventing future attempts to bite.

- **Dogs That Bite**
  - **Statistical Offenses:** The U.S. Centers for Disease Control and Prevention (CDC) has reported the dog breeds most often involved in fatal human attacks (Table 3.5). An association of liability lawyers lists pit bulls, Rottweilers, chows, and Akitas as the most dangerous dogs. In 2009, the U.S. Army, Air Force, and Marine Corps prohibited pit bulls, Rottweilers, Doberman pinschers, chow-chows, and wolf hybrids in U.S. military housing units due to the risk of severe bites.
  - **How to Avoid Bites and Provocation of Bites** (*Tables 3.6* and *3.7*)
  - **Additional Information:** Visit: http://dogbitelaw.com/ and http://dogsbite.org
- **Adult Handler Response to Dog Attack**
  - **Be Proactive, Not Reactive:** Defense in dog attacks can be either reactive or proactive, but preparation for both is advisable (*Tables 3.8–10*). Basic defense includes not screaming, avoiding eye contact, remaining motionless, and backing away slowly when the dog moves away or hesitates.

Table 3.5	Dogs Most Involved in Fatal Human Attacks in the U.S., in descending order
•	Pit bulls
•	Rottweilers
•	German shepherd dogs
•	Huskies
•	Alaskan Malamutes
•	Doberman pinschers
•	Chow-chows
•	Great Danes
•	Akitas

Table 3.6	Avoidance of Dog Bites
•	Select a dog appropriate for a family's living situation and family members.
•	Socialize the dog to other humans and animals in the first 4 months of its life.
•	Train the dog to obey simple commands.
•	Keep the dog on a leash in public.
•	Avoid aggressive games like wrestling or tug-of-war with the dog.
•	If the dog is male, it should be neutered early in life.
•	Provide and maintain a safe and secure containment (well-maintained fence, kennel, crates) and strictly avoid tethering.
•	Give each dog playtime and short periods of training each day.
•	Avoid actions that provoke dog attacks.

Table 3.7 Actions that can Provoke Dog Attacks

•	Tethering
•	Teasing or taunting
•	Invading a dog's territory
•	Wrestling games with dogs
•	Loud, sudden noises (firecrackers, gunfire)
•	Competing for food (bothering a dog that is eating)
•	Presence of a female dog in heat
•	Demonstrating fear

Table 3. 8 Proactive Defense to Dog Attack

•	Never try to handle an aggressive large dog without another capable handler present.
•	The nonlethal dog defense weapons policy of the American Veterinary Medical Association is that electromuscular disruption devices (EMDDs), also called stun guns or tasers, should not be used on any animal for routine capture or restraint.
•	Sprays that use capsaicin, citronella, and similar irritants require close proximity and accurate aim and can infuriate an excited aggressive dog rather than deter it.
•	Air horns can be effective deterrents at a greater distance, they do not require aim, and they can deter multiple dogs simultaneously as well as alerting others to either help or to avoid the aggressive dog.
•	Another proactive defense against dog bites is to encourage the elimination of bite-provoking stimuli (Table 3.9).

Table 3.9 Dog Bite-Provoking Stimuli

•	Fleeing as if prey
•	Tethering a dog on a rope or chain
•	Teasing or taunting dogs
•	Play wrestling with dogs
•	Perceived need to protect food or puppies
•	Presence of a female dog in heat
•	Loud noises such as firecrackers or gunfire

Table 3.10 Preparation for an Impending Attack by a Large Dog

•	*Avoid running from the dog* if not absolutely positive there is time and a definitive way to escape.
•	An obstruction (bag, backpack, umbrella, coat, bicycle, car, etc.) should be sought to be between the victim and the dog.
•	Order the dog to *BACK OFF* with a low, stern voice and occasional yells for help.
•	Wrap an arm with a coat can help in fending off an attack, and if the dog attacks an arm, kick it hard and repeatedly until it releases.
•	A nearby stout stick or similar object should be sought that can be used to keep the dog at bay while backing toward safety.
•	If there is no escape evident and no nearby object to use as a weapon, a stationary object should be grabbed to prevent the dog from knocking or pulling a victim to the ground.
•	If knocked to the ground, a victim should curl up in fetal position and press his fists into his neck while keeping his elbows firmly against his chest and his legs curled up and held tightly together.

DOG SAFETY

More dogs are subjected to pain and suffering by poor care by humans than there are humans who are seriously bitten by dogs.

- **Proper Puppy Socialization Is Critical:** Lack of socialization and improper containment are the primary safety hazards to dogs. Many problems that lead to relinquishing dogs to shelters stem from poor socialization while a puppy.
- **Improper Containment:** Improper containment such as tethering or allowing dogs to ride loose in pickup beds, campers, or house trailers puts dogs at multiple risks. A lack of containment can lead to dogs being hit by cars, running in packs, and other harmful sequelae.
- **Dog-to-Dog Interactions:** Improper introduction of a dog to new dogs can be hazardous (*Procedural Steps 3.2*).

Procedural Steps 3.2	Proper Introduction of a Dog to New Dogs
1.	Distractions should be minimized.
2.	The introduction should be on neutral ground with both dogs on a short leash, preferably with just one handler per dog.
3.	Time for the dogs to assess each other by sight and smell at a distance from each other is important.
4.	Based on the body language of each, the distance can be gradually reduced until they can do anogenital smelling.
5.	Signs of overstimulation and the need to separate dogs include growling, teeth baring, prolonged direct stares, stiff-legged gait, and attempting to stand on top of the other dog.
6.	If it is necessary to separate dogs, each handler should pull them apart using the dogs' leashes or pressing a panel between them.
7.	*Note:* Trying to grasp the dogs, even by their hind legs, to pull them apart is dangerous for handlers.

KEY ZOONOSES

Apparently healthy domestic dogs pose little risk of transmitting disease to healthy adult handlers who practice conventional personal hygiene. The risks of physical injury are greater than the risks of acquiring an infectious disease.

Directly transmitted zoonotic diseases from dogs can result in signs of disease systemically or primarily in the respiratory, digestive, or integumentary system of humans. In some cases, healthy-appearing dogs can transmit zoonotic diseases (Table 3.11).

SANITARY PRACTICES

- **Zoonosis Prevention**
  - **Dress, Vaccinations, Control of Parasites**
    - **Wear Proper Apparel:** A handler of dogs should wear appropriate dress to protect against skin contamination with hair and skin scales or saliva, urine, and other body secretions.
    - **Avoid Dog Saliva:** Handlers should not allow dogs to lick their face, wounds, or scratches.
    - **Control Ectoparasites:** Fleas, ticks, deerflies, and other biting flies should be controlled.
    - **Keep Vaccinations Current:** Vaccinations in dogs should be kept current against rabies and leptospirosis. Dogs should be dewormed on a routine conventional schedule. Dog handlers should be vaccinated against tetanus at least every 10 years.

Table 3.11 Diseases Transmitted from Healthy-Appearing Dogs to Healthy Adult Humans

Disease	Agent	Means of Transmission	Signs and Symptoms in Humans	Frequency in Animals	Risk Group*
Bites	—	Direct injury	Bite wounds to face, arms, and legs	All dogs are capable of inflicting bite wounds	3
Visceral larvae migrans	*Toxocara canis*	Direct, fecal-oral	Enlarged liver, coughing	Very common in puppies	2
Ocular larvae migrans	*Toxocara canis*	Direct, fecal-oral	Defects in eye(s), blindness	Very common in puppies	2
Cutaneous larvae migrans	*Ancylostoma caninum, A. braziliense, Uncinaria stenocephala*	Direct, oral, transdermal	Linear, red, itchy eruptions on the skin	Common in puppies	2
Echinococcosis	*Echinococcosis granulosus, E. multilocularis*	Direct: fecal-oral	Cough and shortness of breath, or abdominal pain and jaundice	Possible in dogs allowed to eat sheep or rodents	4
Leptospirosis	*Leptospira* spp.	Direct, oral, mucous membranes, broken skin	Flu-like signs and symptoms	Common if dogs are near wildlife	3
Campylobacteriosis	*Campylobacter* spp.	Direct, fecal-oral	Diarrhea	Common in puppies	3

*Risk Groups (National Institutes of Health and World Health Organization criteria. Centers for Disease Control and Prevention, Biosafety in Microbiological and Biomedical Laboratories, 5th edition, 2009.)

1: Agent not associated with disease in healthy adult humans.

2: Agent rarely causes serious disease, and preventions or therapy possible.

3: Agent can cause serious or lethal disease, and preventions or therapy possible.

4: Agent can cause serious or lethal disease, and preventions or therapy are not usually available.

- **Practice Basic Sanitation**
  - **Clean Hands:** Keep hands away from eyes, nose, and mouth when handling dogs, and wash hands after handling them. Handlers should wash their hands each time they handle pet foods and treats.
  - **Avoid Fecal Contamination:** Feces should be removed from yards and properly disposed of at least weekly. Dogs should be prevented from eating out of cat litter boxes.
  - **Clean Food and Water Bowls:** Dog food bowls and food scoops should be washed after each use.
  - **Prevent Feeding on Raw Meat:** Dogs should not be allowed around rabbit or rodent burrows or given the chance to kill or eat dead wild rabbits or rodents. The Federal Drug Administration and Centers for Disease Control discourages the feeding of raw meat or bones to dogs due to the risks of transmitting salmonellosis, listeriosis, and colibacillosis.
- **Avoid Fleas and Ticks:**
  - **Fleas and Ticks on Dogs:** Children should not handle dogs with fleas or ticks. Dogs should be routinely examined and treated for external parasites.
  - **Ticks in the Environment:** Avoiding ticks requires avoiding tall grass, brush, and bushes during warm weather and keeping grass short in dog pens and yards. Wearing light colored clothing facilitates seeing and removing ticks. Long sleeves and long pants that are tucked into socks plus tall boots and a hat reduce the possible tick attachment sites.
  - **Attachment Deterrents:** Skin and clothing can be treated with N, N-diethyl-m-toluamide (DEET), or just clothing can be treated with permethrin to deter or kill ticks.
  - **Daily Inspections:** Daily inspection of the skin, particularly under long hair of children, and prompt removal of attached ticks will minimize or eliminate the risk of transmission of zoonotic diseases. Most tick diseases take 24 to 48 hours of attachment for disease transmission to occur.
- **Preventing Spread of Disease among Dogs**
  - **Initial Control Measures:** When handling more than one dog from different households or kennels, proper sanitation is required to prevent the spread of disease from carriers without clinical signs to dogs immunologically naive to the disease.
    - **Isolate New Dogs:** Dogs from different origins should not be confined in the same cage or run. A separation of at least three feet is desirable to reduce the risk of the spread of airborne disease agents.
    - **Maintain Sanitation:** Handlers should wash their hands before and after handling animals and clean and disinfect tabletops and cages used in handling. Runs should be sanitized with chlorine (3 cups of bleach/gallon of water) before being used by a dog that has not previously mingled with other dogs that have used the run.
    - **Sanitize Potential Fomites:** Restraint equipment such as blankets, muzzles, capture poles, grooming equipment, collars, harness, and slip leashes should be disposable or cleaned and disinfected. Leather gloves should be kept as clean as possible and used infrequently.
  - **Sick Dogs:** Special precautions are needed if sick dogs are handled, and sick dogs should be isolated from apparently normal dogs. New household dogs should be quarantined for at least 2 weeks to reduce the risk of transmitting a disease to other dogs in the house.

Whenever possible, handlers should allow a dog the opportunity to approach and be caught rather than the handler approaching a dog to catch it (Table 3.12). If the owner is present, the handler should first speak with the owner and initially ignore the dog. This allows the dog to assess the handler's voice, body language, and acceptance by its owner.

COMPANION AND WORKING DOGS

- **Initial Approach**
  - **Indirect Observation:** The dog's attitude should be observed to determine if it appears friendly and calm (typical of most companion dogs); friendly and fearful, fearful and reclusive or aggressive, or dominance aggressive. The handler should avoid a fixed stare or staring at the dog's eyes.
  - **Calm Voice:** A normal quiet, managed tone with reassurance should be used. The dog should be called by its name, if known, when speaking to it. A quiet, cheerful tone should be used, and an overly excited, party-time voice should be avoided.
  - **Lower Your Body:** Kneeling when attempting to first interact with a strange dog is less threatening to dogs. However, a handler should never sit on a floor when handling a strange dog or one with a history or body language of aggressiveness.
- **Respect Personal Space:**
  - **Avoid Entrapment:** Dogs should not be approached or attempted to be caught in a small, confined space. In a relatively open area, the dog should only be approached up to the edge of the dog's personal space zone (usually about 3 feet).
  - **Avoid Requiring a Direct Approach:**
    - **Turn and Crouch:** The handler then should stand sideways or crouch with his side to the dog and give it a chance to more easily approach submissively. If the dog is large and potentially aggressive, the handler should be positioned so that he can stand immediately and move if needed.
    - **Avoid Intimidating Behavior:** Greeting an unfamiliar dog should NOT involve a direct confrontation, leaning over the dog, patting on top of the dog's head, thrusting a hand with outstretched fingers in front of it, a high, squeaky voice, or direct stares.
- **Use Positive Reinforcement:**
  - **Food Treats:**
    - **Presentation:** Food treats may be held out at the level of the dog's head or tossed near the dog to entice it to approach.

Table 3.12 Proper Approach to Dogs Without Signs of Aggression	
•	Observe casually; no staring.
•	Use a calm voice and calm (slow, deliberate) movements.
•	Lower your body.
•	Avoid entrapment.
•	Use small treat as a lure when allowed.
•	Avoid leaning over dog's body and reaching on top of dog's head.
•	Use short praise, when deserved, and avoid constant praise.

- **Types:** The treat should be small and easily consumed in a couple of seconds. Dry dog food treats that are easily stored in a pocket and will not spoil are best. However, some handlers prefer to use pieces of boiled hot dog, dried shrimp, or canned cheese spread.
  - **Avoid Constant Praise:** Constant praise should not be used for the dog's approach. Praise should be metered out and appropriate to each stage of the behavior to be effective.
- **First Interaction—Petting**
  - **Protect Fingers:** After the dog has approached the handler, the handler should offer the back of his hand with his fingers curled for the dog to sniff. The hand should be offered at the level of the dog's head or lower. A possibly fearful or otherwise aggressive dog should never be approached by offering a hand with extended fingers to smell.
  - **Stroke Side of Face:** If the dog's body is relaxed and the dog sniffs or licks the hand, it can then be stroked on the jaw or side of the face. Petting should not be initially directed toward the top of the dog's head or shoulders, and the dog should not be leaned over. The rest of the dog's body should be gently stroked from the neck toward the hips before attempting to move or lift it.
- **Second Interaction—Slip Leash**
  - **Apply Leash:** Once the dog tolerates being petted, a slip leash should be placed over its head and around its neck. When possible, the leash loop should go around the neck and one front leg on small dogs.
  - **Separation from Owner:** When separating a dog from its owner is needed, the owner should move away from the dog whenever possible rather than leading or carrying a dog from its owner. Large dogs can be led, and small dogs are usually picked up.
  - **Assess History of Pain:** If an owner is present, he should first be asked if the dog is known to be painful anywhere before the dog is picked up.
  - **Lift Head Forward with Leash:** The slip leash is gently pulled forward and upward for head restraint while the other hand reaches under the dog and supports its body to be picked up. The hand with the leash can then be moved to the dog's neck to aid in support and loose control of the head (Figure 3.2).
- **Complication of More Than One Dog:**
  - **Address Dominant Dog:** If there is more than one dog, the dominant dog should be addressed first and control of it established before proceeding to other dogs.

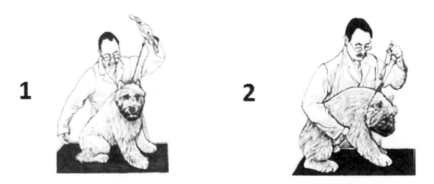

Figure 3.2 Lifting slip leash to control head movement for capture.

- **Ask to Sit:** Most companion dogs know the command to sit. If the dog is fearful or overtly aggressive, it should be given the command to sit. Whether it complies or not, can be an indicator of its continued apprehension or aggression and the need for greater physical restraint methods or for chemical restraint.
- **Enlist Assistant When Appropriate:** Large potentially dangerous dogs should never be handled by one handler alone.
- **Potential Aggression**
  - **Breeds:** Some dogs, particularly retrievers, herding dogs, and guard dogs, are more aggressive or defensive when the owner is nearby. Other dogs, such as terriers, may be more difficult to handle when the owner is gone.
  - **Novices Not Allowed:** If a dog has a history of biting or obviously is aggressive, an owner or any other nonprofessional handler should not assist in the dog's capture or restraint.
- **Escaped Dogs**
  - **Avoid Chase:** Capture of an escaped dog should not involve chasing it.
  - **Approach and Catch:** Once located, the dog should be approached slowly to the closest distance that does not appear to threaten the dog. The handler should kneel, speak calmly, and offer small food treats until it approaches and permits petting.
  - **Using a Lure:** Having a friendly dog on a leash accompany the handler can be an added lure when attempting to capture an escaped dog.
  - **Humane Capture Cage:** Humane capture cages commonly used to harmlessly trap raccoons and other nuisance wildlife may be used to capture escaped dogs that cannot be caught directly.

SERVICE DOGS

- **Service Dogs and the ADA:** Service dogs, as defined by the Americans with Disabilities Act (ADA), are dogs that have been trained to assist people with disabilities. Their training can be for up to 2 years, and the cost may exceed $40,000. Unfortunately, in the U.S., no certification of training of service dogs is required by the ADA.
- **Verification of a Service Dog**
  - **No Vest or Tag Required:** Service dogs are not required by the ADA to wear a vest or ID tags.
  - **Legally Allowed Questions:** The ADA only permits two questions to be asked of an owner of a possible service dog (Table 3.13).
- **Conduct Around a Service Dog:**
  - **Do Not Distract:** When a service dog is in a harness or on a leash, it is working and should not be distracted. It should not be talked to except by the

Table 3.13	ADA Permitted and Disallowed Questions of Owners of a Service Dog
Allowed:	
•	Is the dog required because of a disability?
•	What assistance has the dog been trained to do?
Disallowed:	
•	Questions may not be asked of the person's type of disability or for the dog to demonstrate its ability to assist.
•	Documentation of a person's disability or the dog's training cannot be requested.

owner or petted except by the owner. The owner should not be distracted. Discussion with the owner must wait until he or she appears free to talk.

- **No Treats:** The dog should not be offered treats or snacks by anyone other than the owner.

- **Emotional Support Animals:**
  - **Typical Certification Process:** Certification as an emotional support animal is currently available online for a fee and filling out a brief questionnaire. The qualifications for being a *mental health professional* are not well defined.
  - **Mental Health Professionals:** Psychologists who certify animals as *emotional support animals* do so without protection by the ADA. Mental health professionals could also face legal ramifications for such actions because of the lack of scientific guidelines and the risk of the animal causing disease or injury to others. It is permissible to ask if the owner has a letter from a mental health professional stating the requirement for an emotional support animal, but you cannot ask to see it.
  - **Forensic Psychologist:** True emotional support animals should be certified by a forensic psychologist who is capable of providing legal defense service if needed.
- **Assistance Animals:** Assistance animals are a broader definition and do not fall under the ADA, Most countries maintain a clear distinction between the designations of a service dog and an assistance dog. Also, owner training of service dogs is not permitted in some countries.

HANDLING FOR ROUTINE CARE AND MANAGEMENT

BASIC EQUIPMENT

When handling a dog other than the handler's personal pet, a slip leash is the most useful equipment. All dogs (and cats) in a veterinary clinic or boarding kennel should have a slip leash on when taken outside a cage or kennel. All dogs taken outside a building without a secondary barrier to escape (fence) should have a chest harness with an attached leash.

- **Leashes**
  - **Flat Training Leashes with Metal Snaps:** Flat leashes with metal snaps to attach to metal rings on collars are in common use, in part, because the flat strap provides a surface for marketing words or figures from the seller. Snap-on leashes and buckle, flat collars provide little restraint, easy escapes by motivated dogs, and are poor training tools. Snap leashes should be attached only to a slip-chain training collar, head collar, or chest harness.
  - **Slip Leashes**
    - **Definition:** A slip leash is a rope, cord, or flat woven strap with a metal ring honda or tied honda knot used for routine handling of dogs. Flat-strap slip leashes should not be used due to their inability to maintain an open loop when being placed over the dog's head and neck.
    - **Purpose:** A slip leash serves as a sliding collar and lead rope in one piece. It can be tightened when needed to gain the dog's attention and released to reward proper responses. It also provides greater security against escape than a fixed collar and snap leash.
    - **Application:** The handler should not stand in front and extend his hands toward the dog to place a slip leash. This posture is intimidating to dogs.
    - **Orientation of Neck Loop:** Handlers traditionally stand or walk with the dog on the handler's left side. For the slip leash to loosen when desired

properly the honda end of the leash should go clockwise around the dog's neck. This allows the neck loop to loosen when tension is released on the leash.

- **Fractious Dogs:** When dealing with fractious dogs, a string should be tied to the leash's honda. The slip leash can then be loosened by pulling on the string and removed without placing a hand near the dog's head.
- **Contraindication:** Slip leashes should not be used on dogs with breathing problems. If an alternative does not exist, the loop should be placed around the neck with one front leg through it to prevent pressure on the trachea. Dogs should never be tied and left unattended with a slip leash.
- **Misapplication as a Muzzle:** Slip leashes should not be wrapped around the muzzle and held in place by the ends to form a temporary muzzle. The neck loop could be too tight when the loops around the muzzle are made. Use of a slip leash as a muzzle can also create an aversion to any use of slip leashes on the dog.
- **Retractable Leashes:** Training leashes are 4 to 6 feet long. A retractable leash is a snap band or cord leash that is 10 to 26 feet long and can be spring wound, similar to the action of a retractable measuring tape. Retractable leashes offer minimal control of dogs and should only be used in open spaces.
- **Collars:** No collar will inherently injure dogs, while any collar can injure dogs if a handler applies inappropriate force to the collar.
  - **Flat and Rolled (Fixed-Buckle) Collars**
    - **Uses:** Flat fabric or flat or rolled leather collars with a buckle (Figure 3.3) or plastic snap closure (Figure 3.4) is used for identification purposes and routine restraint of puppies or sensitive small dogs. Collars should allow 2 fingers to be easily slipped underneath, or the collar is too tight.

Figure 3.3 Flat collar with buckle closure.

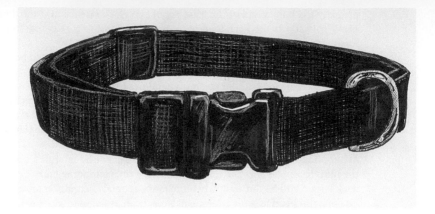

Figure 3.4 Flat collar with snap closure.

Figure 3.5 Slip-chain training collar.

- **Risk of Escape:** Leashes should not be attached to flat or rolled collars for reliable restraint. Even if the collar is properly fitted, the dog may be able to back up, shake its head, and escape.
- **Slip-Chain Training (*Choke*) Collars**
  - **Misleading Name:** Slip-chain training collars are similar to slip leashes (Figure 3.5). The term *choke* is a misnomer since the goal is not to choke the dog.

- **Collar Action and Proper Use:** A training collar tightens quickly around the neck and releases quickly when tension is released on an attached leash. Pulls should be to the side and not upwards which can cause excessive compression around the neck.
- **Orientation of Neck Loop:** For the handler to be on the right side of the dog's body, the collar's loop should go clockwise around the dog's neck. If applied counter clockwise, the loop will not fully release when tension is removed on the attached leash. When in doubt, place the collar on the left wrist and pull upward then release to see if the chain relaxes as it should (Figure 3.6).

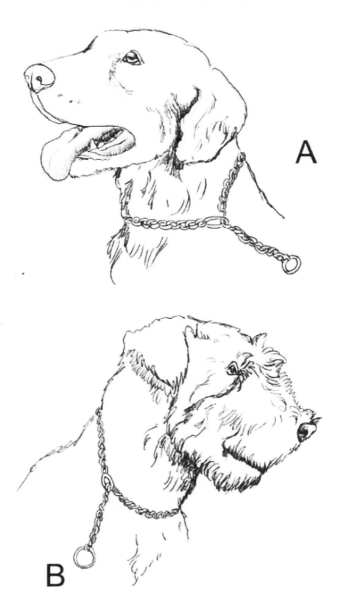

Figure 3.6 Proper placement of a slip-chain training collar. A is proper and allows release. B is improper and does not allow release.

- **Dogs Walk on Left Side:** Dogs must continually walk on the same, traditionally left, side of the handler. The chain will not be oriented correctly when on the other side of the handler.
- **Training Use Only:** Training collars should only be used for training purposes and when the dog is on a hand-held leash. Otherwise, there is risk of strangulation if tied or if the collar becomes caught on an object.
- **Misuse Potential:** Slip-chain training collars, as with any restraint equipment, can be misused and cause aversive behaviors. Unlike shock collars, slip-chain collars are training tools that can deliver attention-to-the-handler signals appropriate to situations that may quickly vary. Used with proper discretion and timing, a slip-chain training collar can be a safe, useful communication tool between handlers and dogs and does not cause aversive behaviors.

● **Martingale and Prong Collars**
  - **Description:** Martingale collars are flat collars with rings at both ends and a chain that goes through each ring (Figure 3.7). The chain also has rings at each end. The size of the flat collar can be adjusted in size so that the extent of squeeze on the neck when the leash is pulled can be modified.
  - **Limited Slip:** Martingale collars are also called limited slip collars because they are less likely to slip off if the dog pulls back on the collar and leash. For this reason, they are often used on sighthounds, such as Afghans, which have narrow heads.
  - **Comparison to Slip-Chain Collars:** Unlike slip-chain collars, martingale collars cannot be put on backward and will work the same if the dog changes from one side of the handler to the other. Since they are flat collars and the pressure delivered to the dog's neck from tension on the leash is less than a slip-chain collar, martingale collars can be more easily ignored than slip-chain collars by inattentive dogs.
  - **Blunt Prongs:** Thick-haired dogs may have sufficient hair padding on their neck to ignore the pressure of a basic martingale collar. Blunted prongs can produce better responsiveness without causing injury (Figure 3.8).

Figure 3.7 Martingale collar.

Figure 3.8   Blunt-prong martingale collar.

- **Chest Harness**
  - **Primary Advantages:** Chest harnesses cannot strangle dogs or aggravate upper respiratory disorders. They will not slip off if the dog pulls backward on a leash. All dogs without advanced leash or voice command training taken outside a building or security enclosure should have a harness with attached leash on to prevent escape.
  - **Control of Vicious Dogs:** Vicious dogs may be more easily handled and tractable by wearing both a harness and a collar with leash that cannot be chewed in two (chain or heavy wire).
  - **Front Clip:** Some chest harnesses have a front clip for a leash. A restraining pull from a leash to a front collar disengages the pulling power of a dog and may aid in discouraging pulling in the future. A front-clip chest harness is less effective than a head halter but easier to use, and dogs adjust quicker to the chest harness than a head halter.
- **Towels and Blankets**
  - **Less Anxiety for Dogs:** A towel or blanket that a dog is familiar with and has the dog's or owner's scent on it can be comforting to a dog and reduce its fear when handled.
  - **Safety Tool for Handlers:** Towels or blankets can be used to cover the dog's head to facilitate grasping the neck for head restraint. They can also be rolled into a bulky soft collar to go around the neck for mild restraint of the head. An aggressive dog can be distracted with towels or blankets and allowed to bite them while the handler's other hand approaches from the rear to capture and restrain the dog.
  - **Do Not Towel and Scruff:** When using a towel or blanket over a dog's head, the scruff hold should not be attempted. It is safer and more effective to grasp both sides of the neck (two hands neck hold) just behind the ears with thumbs on the back of the head and fingers underneath the mandible.

- **Tables and Table Covers**
  - **Importance of Tables for Handling:** Tables that place a dog at the handler's waist height will eliminate the leaning position, which is intimidating to dogs, and facilitates a detailed physical examination. Also, a proper physical exam CANNOT be performed on a dog under 50 lbs. when it is on a floor, bench, ottoman, etc. or while any dog is in a car, truck, crate, etc.
  - **Slick Table Tops:** A slick table top, in addition to the height, reduces most dogs' desire to escape. Tables with surfaces that provide traction can embolden some dogs to struggle to jump off that would not try otherwise. Slick-top tables are also easier to clean and disinfect and therefore best for general use.
  - **All Types Should Remain Sanitary:** Some nonaggressive dogs may feel too insecure on a slick table and need a washable pad on the table that provides traction, insulation, and can be easily sanitized after use. Sanitation should never be compromised by using a table surface that provides traction and warmth but cannot be sanitized after each use.
  - **Constant Supervision Required:** Whenever a dog is on a table, someone's hand or hands must always be on it to prevent it from trying to jump off.
    - **No Jumping On or Off:** Dogs should never be encouraged to jump onto or off of an exam or grooming table. Jumping off a table will encourage future attempts to jump.
    - **Steps and Ramps:** If steps or a ramp are used to allow dogs to walk up onto tables, the surface of the steps or ramp should be skid-proof but easily sanitized.
    - **Grooming Tables:** Grooming tables with a grooming arm (a table attachment for a leash) should have a neck loop and quick release to prevent strangulation if the dog falls or jumps off the table.
  - **Stability Critical:** A table that can tip over should never be used. Tables with four corner legs are much more stable than a single pedestal table or scissors-action variable height table. Pedestal tables should be bolted to the floor to prevent tipping with heavy dogs when the dog is not properly centered on the table.
- **Muzzles**
  - **Pros and Cons:** Although muzzles can provide a degree of safety from being bitten by a dog during handling, the use of muzzles on dogs can make dogs more fearful of handling. They can be dangerous to the dog when used on older dogs or dogs with respiratory or digestive problems. Muzzles should be used selectively and not as standard policy.
  - **Indications:** Muzzles should only be used on specific dogs with a reasonable need for safety (Table 3.14)
  - **Safety Advantage Limited:** Muzzles will not prevent a handler from being injured by a dog. Dogs wearing a muzzle can cause painful injury by bruising the bones of the hands or face while attempting to bite a handler. Fortunately, dogs often become more submissive and easy to handle if a muzzle is applied.
  - **Safety for Muzzled Dogs:**
    - **Must Be Sanitized:** Muzzles used between dogs can be a highly effective fomite (object that transmits disease). Muzzles should be clean, sanitary, and smooth where they touch the dog's face.

Table 3.14	Indications for Using a Muzzle on a Dog
•	A history of biting
•	Clear indications of aggressive or defensive biting
•	Severe injury or illness with pain
•	No current rabies vaccination

- – **Must Be Worn Properly:** They can also cause injury to the dog when improperly used. It should be determined that the fasteners work easily before attempting to use a muzzle. A muzzle should not impinge on the dog's eyes.
  - – **Guard Against Self-Injury:** The dog should not be allowed to paw at the muzzle, as injury to the face or removal of the muzzle may occur.
- **Types:**
  - – **Definitions:** Styles are open-ended (*sleeve*) or closed-ended (basket muzzles).
  - – **Open-Ended Muzzles:** Open-end muzzles keep the mouth from opening any farther than to be able to lick (Figure 3.9). Panting or drinking water is not possible.
  - – **Closed-Ended Muzzles:** Basket muzzles are closed on the end and allow the mouth to open (Figure 3.10). They are made of plastic or wire. Basket muzzles allow dogs to pant, drink water, and be given small treats for positive reinforcement.
- **Cautions and Contraindications:**
  - – **Respiratory Distress and Inhalation:** Dogs that have recently vomited or have respiratory distress should not be muzzled. If they vomit, the vomitus will be inhaled into the lungs and can cause fatal pneumonia. Brachycephalics (short-nosed) dogs are better restrained by a rolled towel around the neck and behind the ears than by a muzzle.
  - – **Overheating Risk:** Dogs with an open-ended muzzle cannot pant and therefore cannot cool their bodies if their mouth is held shut, and they will overheat.
  - – **Requires Constant Monitoring:** A muzzle should never be left on longer than necessary for handling. Open-ended muzzles should never be left on for more than 15 minutes and never used if the dog will be unattended with the muzzle on.

Figure 3.9   Open-ended muzzle.

- **Commercial Muzzles**:
    - **Advantages:** Commercial muzzles are strong, preshaped, and easy to apply and fasten.
    - **Disadvantages:** Their disadvantages are cost, difficulty in sanitizing, and the need for multiple sizes if many types or ages of dogs are being handled.
    - **Construction:** Commercial muzzles are made of leather, wire, plastic, or nylon. Leather, plastic, and wire muzzles go on more easily than nylon because nonfabric muzzles maintain their shape.
    - **Buckles or Snaps:** Leather muzzles are fastened by buckles, which are relatively slow to fasten. Cloth (fabric) muzzles are often fastened by a belt snap, which are faster than buckles but cause a snap noise near the dog's ear when fastened.
    - **Sanitation of Plastic, Nylon, and Wire Muzzles:** Muzzles should be cleaned and, if possible, sanitized prior to each use. Plastic, nylon, and wire muzzles can be sanitized with common disinfectants.
    - **Sanitation of Leather Muzzles:** Leather muzzles cannot be easily sanitized. Untreated leather muzzles are porous and can trap microorganisms. Leather will also dry out and crack without proper care (Table 3.15).
    - **Application of Muzzles (*Procedural Steps 3.3 and 3.4*)**

Procedural Steps 3.3	Application Method on Tractable Dogs (Figure 3.11)
1.	Tractable dogs or dogs that have been trained to accept a muzzle can be muzzled by one person with a commercial muzzle from behind.
2.	The muzzle straps are held in each hand with the muzzle below the dog's throat.
3.	The muzzle is then quickly and smoothly brought up and over the dog's muzzle.
4.	*Note:* Approaching the dog from directly in front of its nose with a muzzle will cause most dogs to resist.

Procedural Steps 3.4	Application Method on Resistant Dogs
1.	If the dog is not trained to accept a muzzle, commercial muzzles are best applied by two handlers.
2.	One (restraint) handler should have the dog restrained in the sitting position or in sternal recumbency, and the other (muzzling) handler approaches from the side or behind.
3.	The restraint handler's thumbs are positioned behind the dog's ears, palms restraining side movement of the neck, and index and middle fingers beneath the jaws to keep the jaw from being lowered.

Table 3.15	Care of Leather Muzzles
Initial Treatment:	
•	Before their first use, leather muzzles should be treated with neatsfoot oil, dried, and then rubbed with a beeswax-for-leather treatment.
•	This treatment will prevent cracking and inhibit absorption of microorganisms.
•	It will also permit rinsing and drying between each application of the muzzle.
Maintenance Treatments:	
•	Regular retreatment of the leather with oil and wax is based on the frequency of use of the muzzle, but four times per year should be the minimum.
•	A properly maintained leather muzzle will also become more pliable and comfortable for the animal.

Figure 3.10   Basket muzzle.

4.	If necessary, the hands may be partially protected by leather gloves or a towel.
5.	The muzzling handler stands beside the dog, holding the straps of the muzzle on each side with the muzzle underneath the dog's jaw, and slips the muzzle up and over the dog's nose and fastens the muzzle.

- **Putting Treats in a Muzzle:** Putting treats in a muzzle to encourage acceptance can be dangerous to handlers holding the muzzle in a manner for the dog to explore, dangerous to the dog that may inhale the treat, and is unsanitary, unless the muzzle is thoroughly cleaned and sanitized between each use.
- **Gauze Muzzles:** Nonstretch, 2-inch-wide gauze can be used as a convenient, effective, inexpensive, and sanitary temporary muzzle.
  - **Advantages:** The advantages of gauze muzzles are that they are portable, disposable, inexpensive, soft and noninjurious, and fit all sizes of dogs. A roll of gauze can easily be carried in a pocket and is sanitary, since after a portion is used as a muzzle, the portion can be discarded. Stretchy gauze is not a safe restraint for the handler and should not be used.
  - **Disadvantages:** Disadvantages include that more skill is required to apply a gauze muzzle, and application is slower than with commercial muzzles. Since gauze muzzles hold the mouth closed they can cause an inability to pant or drink. They can also be inappropriately applied so tightly by unskilled handlers that they cause pain and injure the skin around the muzzle.
  - **Application and Removal of a Gauze Muzzle (*Procedural Steps 3.5 and 3.6*)**

Procedural Steps 3.5 Application of a Gauze Muzzle	
1.	Safe application of a gauze muzzle requires two handlers.
2.	One handler restrains the neck and jaw from behind with a two-hand head restraint hold (both hands on the neck, fingers below the jaw, and thumbs behind the ears.)

Procedural Steps 3.5	Application of a Gauze Muzzle
3.	While gripping the head, the handler presses down on the dog's neck and shoulders with wrists and forearms to make it more difficult for the dog to lift a front leg and rake the facial area.
4.	The other handler, who applies the muzzle, stands in front of the dog.
5.	When preparing to apply a gauze muzzle, the length of gauze needed is the length of the handler's arms spread wide apart (about 5 ft.) for small- and medium-sized dogs and twice that distance for large dogs (Figure 3.12).
6.	The handler's hand should not ever get closer than 6 inches to the dog's muzzle while applying the loops and pulling the ties down.
7.	The first loop is made with a double overhand knot, put over the muzzle, and pulled down firmly with a knot on top of the muzzle, because the double overhand knot will spread out the loop, making it easier to get over the muzzle and when pulled down, and it will hold its place better while the second knot is readied.
8.	Another loop is quickly made above the dog's nose with a simple overhand knot, flipped under the muzzle, and pulled down below the dog's lower jaw.
9.	An overhand knot is then made behind the head and under ears and tied with a slip (bow) knot (Figures 3.12 and 3.13).

Figure 3.11   Application of a muzzle to a tractable dog.

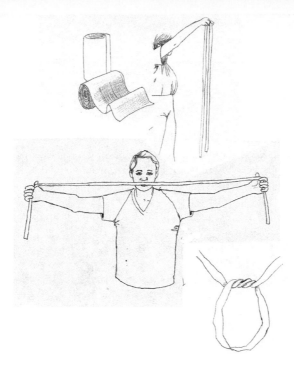

Figure 3.12 Preparing a gauze muzzle.

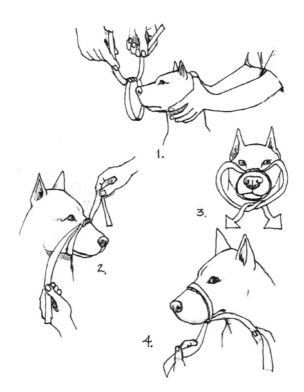

Figure 3.13 Application of a gauze muzzle.

Procedural Steps 3.6	Removing a Gauze Muzzle
1.	To remove a gauze muzzle, the handler's hands should not get closer than 6 inches from the dog's mouth.
2.	The head should be restrained by one handler from behind while the other handler unties the slipknot and then quickly pulls the muzzle in a straight line parallel to the dog's muzzle.
3.	*Note:* The conical shape of a dog's jaws allows the gauze to pull off easily without risking being bitten trying to loosen the ties first.

- **Makeshift Muzzles:** Emergency makeshift muzzles can be created from ties, shoelaces, cords, or long strips of any cloth.

WALKING DOGS

- **Traditional Left of Handler:** Dogs are traditionally walked on the handler's left side. The command for trained dogs is to *heel*. This is an advantage for a handler leading a horse at the same time as a dog, for a right-handed person carrying a hunting rifle, and to position the handler between the dog and traffic when walking along a road on the left side facing traffic. A dog should always be on a short leash when in crowded surroundings.
- **Signal to Begin Walk:** Dogs trained to heel are taught to walk off with the handler if the handler steps off with his left foot, the nearest foot to the dog. If the handler wishes the dog to remain still, the handler steps off with his right foot. Signals to stay in place are done with the handler's left hand, and signals to heel are reinforced with moving the right hand.
- **When Not to Walk:** Whenever small dogs do not follow on a leash, they should be picked up and carried. Dogs should never be dragged by a leash.

LIFTING AND CARRYING DOGS

Lifting of all sizes of dogs from the floor should be done with knees bent and back straight. One person in good health and physically fit can lift dogs up to 50 lbs. For fractious dogs, an assistant may be needed to hold the leash while the handler wraps a rolled towel around the dog's neck or places a muzzle on the dog before lifting it.

- **Small and Medium Dogs Under 50 lbs**.
  - **Small Dogs (*Procedural Steps 3.7*)**

Procedural Steps 3.7	Lifting and Carrying Small Dogs (Less than 35 lbs.)
Lifting	
1.	When picking up a small dog, a slip leash should be applied first.
2.	The leash is pulled forward slightly to prevent the dog from being able to fully turn its head to the side and bite.
3.	The other hand then reaches under the dog's chest and abdomen and supports the body while being lifted.

Carrying	
1.	When carrying a small dog, its head should be held next to the right side of the handler's body.
2.	The left hand is placed under the neck, with fingers on the side of the neck just behind the jaw, and its body is supported with the right forearm and wrist under the dog's chest.
3.	The right hand grasps the dog's left foreleg (the one nearest the handler's body) so that the dog cannot climb up the handler's chest or wiggle away (Figure 3.14).

- **Medium-Sized Dogs (*Procedural Steps 3.8*)**

Procedural Steps 3.8 Lifting and Carrying Medium-Sized Dogs (35–50 lbs.)	
Common Method	
1.	The left arm is wrapped around the front of the dog's chest and under the neck while the handler's right arm is placed around and behind the dog's hind legs.
2.	The left arm can be angled upward on the side of the neck to restrain the head if needed.
Alternative Method	
1.	The dog can be lifted under and around the neck with the left hand and under the abdomen with the right arm, the *forklift* method.
2.	This latter method is the only method that provides control of the dog's head while it is being lifted.

Figure 3.14   Carrying a small dog with restraint of a foreleg.

- **Large Dogs (50–80 lbs.)** (*Procedural Steps 3.9*)

Procedural Steps 3.9 Lifting Large Dogs	
Two Handlers: Two people should lift or carry larger dogs.	
1.	**Front Handler:** One handler restrains the dog's head by his right arm around the dog's neck, and the left arm is placed under and around its chest. The dog's shoulders are pressed against the handler's body, or the handler holds the outside front leg firmly.
2.	**Second Handler:** The second handler lifts the dog's rear portion by grasping both thighs or with an arm under the abdomen and holding the outside thigh.
Lift Tables: Lift tables are also available to assist in lifting a larger dog to table height.	

- **Giant Dogs, Over 80 lbs.:** If necessary, a larger dog can be lifted as with 50- to 80-lb. dogs. However, most procedures on giant breeds are better done on the floor rather than on a table.

HANDLING INVOLVING CAGES AND RUNS

- **Placement of Nonaggressive Dogs in Cages and Runs** (*Procedural Steps 3.10*)

Procedural Steps 3.10 Method for Placement of Nonaggressive Dogs in Cages and Runs	
1.	Dogs should be placed in cages headfirst.
2.	One hand should have control of the cage door.
3.	Closure of the door should begin before release of the dog with the other hand so that when the restraint hand is removed, there is insufficient room for the dog to escape.
4.	Release should be as smooth and quiet as possible, since this will be the predominant memory of being handled.
5.	Removal of a slip lead prior to placing the dog in the cage prevents struggling with the dog in the cage to remove the lead.
6.	*Note:* Dominant aggressive dogs should be kept in lower cages to avoid direct eye contact needed in lifting them and to prevent providing them with a more elevated (dominant) position.

- **Removal of Nonaggressive Dogs from Cages and Runs** (*Procedural Steps 3.11 and 3.12*)

Procedural Steps 3.11 Method for Removal of Nonaggressive Dogs from Cages	
1.	The handler should approach the cage in a friendly manner while speaking to the dog in a calm, cheerful voice.
2.	Removal should begin with using one hand to open the cage door only enough to be able to get the other hand and a slip lead in.
3.	The slip lead is then placed over the dog's head, and the cage door is opened wider.
4.	The dog is assisted by picking it up, or if in a lower cage, leading it out.
5.	When removing a dog from a lower cage, the handler's leg can aid in blocking an escape through the partially opened door while attempting to apply the slip lead.

Procedural Steps 3.12	Method for Removal of Nonaggressive Dogs from Runs
1.	The handler should approach the run in a friendly manner while speaking to the dog in a calm, cheerful voice.
2.	Removal should begin with using one hand to open the run door only enough to be able to get the other hand and a slip lead in.
3.	The handler's leg can aid in blocking an escape through the partially opened door while attempting to apply the slip lead.
4.	The slip lead is then placed over the dog's head.
5.	After the dog's head is controllable with the slip lead, the run door can be opened wider and the dog led out.

TRIMMING NAILS

Dogs that do not frequently walk and run on abrasive surfaces must have their toenails trimmed on a regular basis, generally every 6 weeks. If dog nails touch the floor when the dog is walking, the nails are too long. An important part of puppies' early education should include desensitization to handling of their feet by counterconditioning with food treats.

- **Early Training (*Procedural Steps 3.13*)**

Procedural Steps 3.13	Early Training to Build Tolerance for Nail Trimmings
1.	Use a gentle stepwise process involving a few seconds of handling the upper aspects of each leg and rewarding lack of struggling after each leg has been handled.
2.	Subsequent sessions on following days consist of handling slightly lower aspects of each leg until the foot and nails can be handled without a struggle.
3.	After handling the feet is accepted, one nail on one foot should be trimmed, and the dog should be rewarded with a treat.
4.	The next session should involve trimming two or three nails. (The eventual goal is for the dog to tolerate trimming all nails on all feet in one session. Success may take a couple of days to months.)
5.	Finish each session with a favorite trick, such as *down* or *shake hands*, and a treat.

- **Small Dog Restraint:** Small dogs can be trimmed using the aid of an assistant who holds the dog in their lap. Large dogs should tolerate trimming in a sitting or standing position. Some may roll on their back and lie still while being trimmed.
- **Selective Use of Lateral Recumbency:** Dogs that need immediate trimming to protect them from injury from their long nails and do not tolerate trimming with mild to no restraint can be restrained by an assistant who holds the dog in lateral recumbency (held on their side). However, lateral recumbency should not be a routine restraint for nail trimming.

HANDLING FOR COMMON MEDICAL PROCEDURES

Most handling and restraint of dogs can and should be done without tranquilization, sedation, hypnosis, or anesthesia. However, some handling and restraint procedures may require special skills, equipment, or facilities, and possibly adjunct chemical restraint or complete immobilization by chemical restraint. Owners should be present for routine medical exams since their presence has positive effects on behavior and physiologic measures of fear in most dogs.

RESTRAINT OF INDIVIDUALS OR PORTIONS OF THEIR BODIES

- **Whole Body**
  - **Standing Restraint (*Procedural Steps 3.14*)**

Procedural Steps 3.14	Standing Restraint of Small and Large Dogs
Small Dogs	
1.	The handler stands on the right side of a small dog and restrains the head with the right hand under the throat.
2.	The left hand goes over the dog's back and lifts the abdomen while holding the dog's body next to the handler's body.
3.	The right hand can be used to hold the mouth closed to prevent panting if someone is attempting to auscultate heart or lung sounds.
Large Dogs	
1.	For larger dogs, the handler's right arm wraps underneath and around the neck (the *bear hug* hold) to restrain the head (Figure 3.15).
2.	The left arm is placed under the abdomen to hold the rear end up. The head and abdomen are held close to the handler's body.
3.	When assisting for a rectal exam, the dog's tail is held to keep the dog from sitting while also keeping the tail out of the way of the rectal exam.

Figure 3.15 Standing restraint.

- **Sitting Restraint (*Procedural Steps 3.15*)**

Procedural Steps 3.15	Sitting Restraint of Dogs
1.	The dog is placed in sitting position, and the handler stands on the dog's right side.
2.	The handler's right arm is placed underneath the dog's throat, and the left arm reaches over the flank and holds the dog's body close to the handler's body (Figure 3.16).

- **Sternal Restraint (*Procedural Steps 3.16*)**

Procedural Steps 3.16	Sternal Restraint of Small and Large Dogs
Small Dogs	
1.	With a small dog in a standing position, the handler restrains the dog's head with his right hand.
2.	The left hand is placed on the rump and pushes down gently while tilting the dog's head back and up with the right hand.
3.	After the dog is sitting, the handler reaches around the left side of the dog with the left hand and grasps both front legs.
4.	The dog's front legs are slid forward while the handler pushes the dog's body down with his armpit and chest (Figure 3.17).
Large Dogs	
1.	The handler kneels behind the dog while it is sitting and places his left arm under the neck, and his right hand grasps the front legs.
2.	The front legs are stretched forward while pressing the dog down with the handler's chest.

Figure 3.16   Sitting restraint.

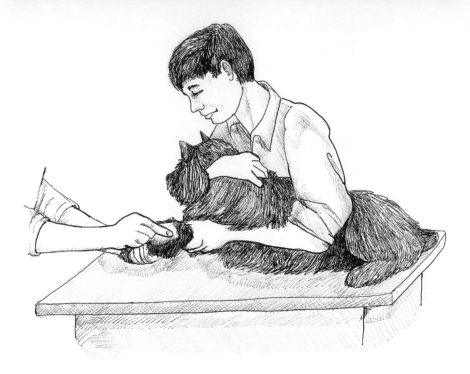

Figure 3.17   Sternal restraint.

- **Lateral Recumbency Restraint** (*Procedural Steps 3.17*)

Caution: Lateral recumbency should only be used if there are good reasons not to use standing, sitting, or sternal recumbency.

Procedural Steps 3.17	Lateral Recumbency Restraint of Dogs
1.	Prior to using this restraint, all needed materials for the procedure requiring lateral recumbency should be made ready to keep the duration of lateral recumbency as short as possible.
2.	The dog should be in a standing position while the handler restrains its head.
3.	For left lateral recumbency, the handler stands on the dog's left side and reaches over its neck with his left hand and grasps the left foreleg.
4.	The handler reaches over the dog's flank in front of the hind leg with his right hand.
5.	The right hand goes underneath the dog's abdomen and grasps the left hind leg.
6.	The dog is lifted while hugging it close to the handler's body so that it gently slides down on its left side, and the handler gently restrains the dog's head with his left forearm and elbow (**Figs. 3.18** and **3.19**).

- **Ventrodorsal Restraint** (*Procedural Steps 3.18*)

Procedural Steps 3.18	Ventrodorsal Restraint of Dogs
1.	A handler may hold the dog ventrodorsal (on its back) to assist with cystocentesis (urine collection by needle and syringe).
2.	A soft, padded surface should be used such as a thick cushion or blanket.
3.	Small dogs can be held ventrodorsal on a handler's lap.

Figure 3.18   Laying a dog in lateral recumbency.

Figure 3.19   Holding a dog in lateral recumbency.

- **Upright Standing Position:** Examination of the lower aspects of the torso or procedures such as cystocentesis may be performed by an assistant who lifts the dog's front legs and puts the dog in an upright stand on its hind feet.
- **Wedging between a Wall and Door:** Aggressive dogs can be restrained for chemical restraint or other injections by pulling their leash through the space between the hinged side of a door and the doorframe with the door partly open (*Procedural Steps 3.19*).

Procedural Steps 3.19	Wedging an Aggressive Dog between a Wall and Door
1.	A door that opens to the inside is opened and the leash is run through the gap between the hinged side of the door and the doorframe.
2.	The dog is pulled toward the angle made by the door and the wall. The door can then swing to push the dog's body next to the wall.
3.	Another handler can then approach the dog's hind quarters for chemical restraint administration, examination of its hindquarters, or intramuscular injection treatment.

- Eye Screw in Wall and Wire Mesh Runs (*Procedural Steps 3.20*)

Procedural Steps 3.20	Restraint of Dogs Using an Eye Screw in a Wall or Wire Mesh Runs
Eye Screw in Wall	
1.	A similar method to wedging with a door can be accomplished anywhere that a large eye screw can be screwed into a wall stud.
2.	The leash is run through the eye screw and the dog's neck pulled up near to the wall.
3.	The dog's body can be pressed to the wall with the handler's leg against the dog's hip.
4.	If necessary, a portable panel (hog or sheep panel) can be used to wedge a dog's body against a wall.
Mesh Wire Runs: Dogs in wire runs can be restrained in a similar manner by pulling their leash through an opening in the wire mesh.	

- Head
  - One-Hand Muzzle Hold (*Procedural Steps 3.21*)
    - **Limitations:** One-hand muzzles can only be applied to dogs that have a long nose and one that a hand can reach around. They are usually used for typical house dogs from 15 to 40 lbs.
    - **Advantages:** This easily sanitized method requires no equipment other than a slip leash, is applied gently and relatively rapidly, the pressure applied is comfortable to the dog, and the release can condition the dog to be less apprehensive when handled again.

Procedural Steps 3.21	Application and Release of a One-Hand Muzzle Hold on Dogs
1.	After placing the dog on a table and applying a slip leash, the neck is lifted slightly.
2.	The handler's nondominant hand scruffs the neck and the slip leash and runs the thumb under the slip leash loop.
3.	The dog's body is pressed against the handler's body.
4.	The handler's dominant hand is placed above the nondominant wrist and while continually pressing the dog toward the handler's body.
5.	The dominant hand is slid over the wrist onto the side of the dog's neck, over the ear, and along the side of the face until the hand is surrounding the dog's muzzle, with the thumb in front of the eyes and the ends of the fingers beneath the jaw.
6.	Constant pressure toward the handler's body by the dominant hand during the movement toward the dog's nose prevents the dog from being able to turn its head and bite.
7.	It is critical that the ends of the fingers of the muzzle hand be pushed into the soft tissue between the jaw bones to prevent the dog from being able to pull back or shake the hold loose.
8.	Release of the one-hand muzzle is a complete controlled reversal of its application, beginning with the nose.

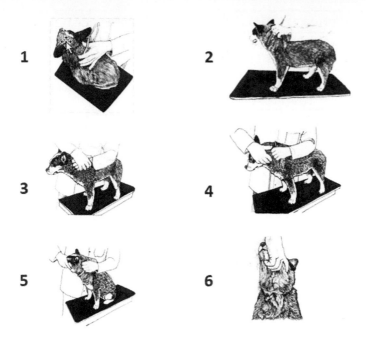

Figure 3.20 One-hand muzzle.

9.	Release should be paused if struggling occurs.
10.	Briefly petting the dog before final release can contribute to counterconditioning it to future handling (Figure 3.20).

- **Two-Hands Neck Hold (*Procedural Steps 3.22*)**

Procedural Steps 3.22	Two-Hands Neck Hold of Dogs
1.	To apply a two-hands neck hold, a handler quickly grasps both sides of the dog's neck just behind the ears.
2.	If the dog is aggressive, placing a towel over the dog's head first may be needed to make it possible to grasp the neck and head without being bitten.
3.	The handler's thumbs should be over the skull, between the dog's ears, and his index finger knuckle should be just behind the angle of the jaw and his fingers under the jaw bones. There should be no pressure on the dog's throat.
4.	Control of side movement of the dog's body is applied with the handler's forearms. Backup is blocked by the handler's body.

- **Mouth**
  - **Restraint of the Mouth Closed:** The one-hand muzzle hold will restrain the head and prevent the dog from opening its mouth.
  - **Restraint to Open the Mouth (*Procedural Steps 3.23*)**

Procedural Steps 3.23	Restraint to Open a Dog's Mouth
1.	To examine inside the mouth, the handler places his nondominant hand on top of the dog's head and muzzle.

Procedural Steps 3.23	Restraint to Open a Dog's Mouth
2.	The dog's cheeks are then pressed between the premolars while carefully putting the index or middle finger of the dominant hand on the lower incisors to entice the dog to open its mouth.
3.	Care must be taken not to let the finger slip off the lower incisors and rake the lower gum.

- **Legs:** A dog's legs can be restrained when the dog is held in the sitting, sternal, or lateral recumbent position. When paired legs are restrained, one finger should be placed between the legs during the hold for comfort of the dog and better traction for the hold.

RESTRAINT OF YOUNG, OLD, OR SICK/INJURED DOGS

- **Puppies:** Bitches should be separated and removed from the room if puppies will be restrained for examination or treatments. During the first 2 weeks of life, puppies cannot see, hear well, or control their body temperature well, so they must be carefully handled and kept warm. Socialization with humans should begin at 2 to 3 weeks of age, but the exertion from handling and length of handling should both be brief.
- **Senior Dogs:** Senior dogs should not be handled for long periods, since they tire easily. Handlers should be mindful of the pain of arthritis common in older dogs. If one is picked up, the dog should be placed on the floor gently.
- **Injured or Sick Dogs**
  - **Pain Aggression:** Normally gentle, friendly dogs will bite if they are in pain. Injured dogs should be muzzled before handling if they are not at risk of vomiting and do not have a head injury or cardiopulmonary distress. Dog crates or cardboard boxes are ideal for transport of small injured or sick dogs.
  - **Stretchers and Slings**
    - **Stretchers:** If a flat, rigid stretcher is used to move a large dog, the dog should be strapped onto it to prevent sliding or crawling off in transit (Figure 3.21). Commercial stretchers for dogs are available with rigid rods attached to the long sides of a sling. This wraps the dog being carried so that it cannot fall off and can be carried more securely with just one hand at each end holding the ends of both rods (Figure 3.22).

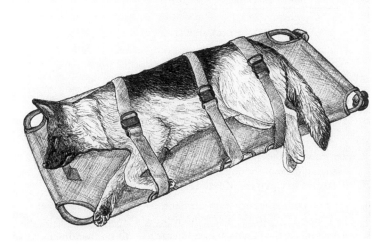

Figure 3.21   Rigid stretcher.

Figure 3.22   Soft stretcher with pole handles.

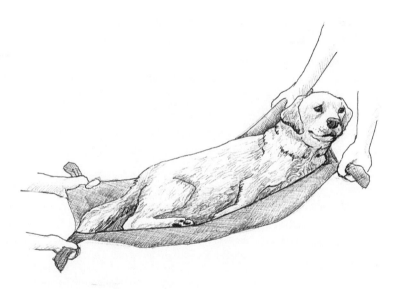

Figure 3.23   Blanket stretcher.

- **Slings:** Two handlers holding four corners of a blanket or large towel can create a substitute stretcher, a sling, which is safer in preventing falls while transporting injured dogs than a flat, rigid stretcher (Figure 3.23).
- **Vertebral Fractures:** Slings do not give adequate support if vertebrae are fractured in the neck or back. Blanket or towel slings require more strength in holding the corners of the fabric than does a rigid stretcher.
- **Gurneys:** Gurneys are tables with locking wheels on the legs are used in veterinary hospitals. Veterinary gurneys are available with which the stretcher can become the top of the gurney. These are scissor-action so that the base can be raised or lowered.

RESTRAINT OF FEARFUL OR AGGRESSIVE DOGS

- **Use Caution and Assistance:** Dogs that show signs of aggression or have a history of biting should never be handled by one handler. Each situation should be evaluated for the best way to minimize stress to the dog, protect the dog from injury, and protect the primary handler and other handlers from injury. This may require towel or blanket restraint, squeeze cage, capture pole, or chemical restraint.
- **Prevention Is Preferable:** Measures should be taken with puppies to prevent fear-related aggression. Use of treats when handling the puppy and socializing it to strange surroundings and different types and ages of people are important aspects of preventing fear aggression.

INJECTIONS AND VENIPUNCTURE

- **Immobilization of Injection Area**
  - **Risks:** Insertion of transcutaneous needles for injection or aspiration in dogs carries the risk of slashing tissue beneath the skin, including damage to nerves and blood vessels, and breaking hypodermic needles off in the dog's body. The area in which the needle is to be inserted must be immobilized, and the dog's mouth and feet should be restrained from interfering with the procedure, especially venipunctures.
  - **Firm, Comfortable Restraint:** The method of restraint should be comfortable, i.e., no squeezing when unnecessary, but the method should allow firm restraint if struggling occurs.
- **Caution: Avoid Topical Anesthetics**
  - **Intended Purpose:** Lidocaine is used topically to numb the skin for some procedures such as tattooing of humans. This has encouraged the use of topical lidocaine in animals to prepare the skin for injections or venipuncture.
  - **Danger:** Dogs, cats, and other animals have the tendency to ingest topical lidocaine, which can cause methemoglobinemia, choking, and an inability to swallow. Topical use of lidocaine in animals is strongly discouraged.
- **Access to Veins**
  - **Definitions:** Venipuncture is also referred to as *phlebotomy*. It is the process of puncturing a vein with a needle and syringe to collect blood samples or to inject medications intravenously. A person who handles the needle and obtains the blood is called a *phlebotomist*.
  - **Flexible Restraint Needed:** Restraint for venipuncture of dogs requires restraints that are comfortable for the dog and do not make it feel trapped, i.e., it should not be squeezed when it is not resisting the restraint. However, the method of restraint should allow the handler to immediately gain tight control if the dog resists.
  - **Sharpest Pain:** The dog will feel a brief sharp pain when the needle goes through the wall of the vein. The dog must not be able to suddenly move, or else the venipuncture may fail.
  - **Veins Must Be Protected:** Dogs with chronic illnesses or those requiring intensive care often have many venipunctures. Ruined venipunctures from improper restraint can result in scarred and thrombosed veins and eventually require surgical cutdowns to achieve access to veins.
  - **Proper Withdrawal:** It is best for the handler and the phlebotomist to coordinate withdrawal of the needle and release of the vein. Release of the vein occlusion should occur before the needle is withdrawn to reduce the leaking of blood from the venipuncture hole. Compression should be applied to the venipuncture site after removal of the needle.
  - **Locations**
    - **Cephalic Venipuncture:** Cephalic veins are on the front of the dog's forearms (*Procedural Steps 3.24*).

Procedural Steps 3.24	Cephalic Venipuncture Procedure in Dogs
1.	The dog should be placed in a sitting position or sternal recumbency facing toward the handler's left side.
2.	The handler's left hand is placed under the dog's neck, and its neck and head are hugged close to the handler's body. (Restraint and proper positioning of the head is important in all cases, since the sting of puncturing the vein wall may cause struggling. Additionally, curiosity makes most dogs hold the muzzle over the leg, obstructing the view of the phlebotomist.)
3.	To restrain a dog for a right cephalic venipuncture, the handler's right arm goes over the dog's body, and the dog's body is hugged close to the handler's side with his elbow while his right hand grasps the dog's outside elbow.
4.	The thumb of the right hand, which holds the elbow, reaches across the front of the elbow, squeezes the leg slightly, and then pulls the skin gently toward the outside (counterclockwise). This occludes and stretches the cephalic vein to facilitate venipuncture by the phlebotomist.
5.	The middle or ring finger of the right hand should be positioned behind the dog's elbow. Otherwise, it is fairly easy for the dog to pull his leg back and out of the handler's grip (Figure 3.24).
6.	Large dogs may be restrained for cephalic venipuncture on the floor, where the dog is put in a sitting position and the handler kneels behind the dog, holding a foreleg for venipuncture with one hand and the dog's head and neck with the other.

- **Lateral Saphenous Venipuncture:** Lateral saphenous veins are on the outside surface of the lower aspect of the hind legs and seen best just above the hock (**Procedural Steps 3.25**).

Figure 3.24   Sternal restraint for cephalic venipuncture.

Procedural Steps 3.25	Lateral Saphenous Venipuncture Procedure in Dogs
1.	The dog must be held in lateral recumbency by the handler for the phlebotomist.
2.	Occlusion of the vein is done by either the phlebotomist or a second handler (assistant).
3.	The assistant needs to use both hands on medium- to large-sized dogs to clasp above the stifle to occlude the vein and prevent the hind leg from flexing, while one hand clasped above the stifle may be sufficient in toy to small breeds.
4.	The foot of the upper hind leg is restrained by the phlebotomist.
5.	Venipuncture of the lateral saphenous vein should NOT be performed with a dog in standing position, since restraint of movement is not possible, and the vein could be badly damaged.

- **Jugular Venipuncture:** Jugular veins are in the lower side of the neck on each side of the windpipe (trachea). Because patent, uninjured jugular veins are extremely important in placing catheters in critical care, jugular venipuncture in dogs should be avoided for nonvital venipunctures, except in extremely small dogs with inaccessible cephalic veins (***Procedural Steps 3.26***).

Procedural Steps 3.26	Jugular Venipuncture Procedure in Dogs
1.	The dog can be restrained in a sitting, sternal, or lateral recumbency position for jugular venipuncture.
2.	Small dogs may be placed in ventrodorsal recumbency (on their backs) for jugular venipuncture.
3.	Large dogs (more than 80 lbs.) are best handled on the floor by putting their hind-quarters in a corner and straddling the trunk of their body or kneeling behind them.
4.	To perform a routine jugular venipuncture in an average-sized dog, the dog is placed on an exam table.
5.	Remove any collars. If a collar is left on and an attempt is made to pull it up and out of the field of the jugular venipuncture, it can compress the neck between the head and the venipuncture site. This will prevent blood from entering the jugular veins.
6.	The dog is moved to the edge of the table and placed in sitting position or sternal recumbency.
7.	With the handler on the left side of the dog, the front legs are held with the handler's left hand.
8.	If the dog is in sternal recumbency, both front legs will have to be held over the edge of the table and downward, and therefore, the edge of the table should be padded with a towel.
9.	The dog's head is held with the handler's right hand on the lower aspect of the jaw, and its head is pointed toward the ceiling.
10.	The phlebotomist occludes the vein.
11.	Movement of the front legs must be controlled, because some dogs will reach up and push the phlebotomist's hand holding the syringe away and ruin the venipuncture and possibly seriously injure the vein. (This is most likely to occur just after the wall of the vein has been penetrated if the front legs are free.)
12.	If the handler cannot hold both front legs with a finger in between the legs for adequate restraint, the handler's arm should be placed around the front of the chest to block the legs from reaching up and pushing the phlebotomist away.

- **Injections**
  - **Subcutaneous (SC):** Dogs are usually held in sitting or standing positions for subcutaneous (beneath the skin) injections by a handler. The person administering the medication or vaccination makes the injection in an upper side of a shoulder area but not on the upper midline.
  - **Intramuscular (IM)**
    - **Immobilization of the Muscle:** Movement of a dog during an IM injection can injure muscle, nerves, and blood vessels. Restraint for intramuscular injection should involve complete immobilization of the area the injection occurs in.
    - **Restraint Positions:** The dog may be restrained in standing or lateral recumbency positions.
    - **Injection Locations:** Intramuscular injections can be given into the back of the thigh muscles (semimembranosus and semitendinosus), front of the thigh muscles (quadriceps), muscles above the elbow (triceps), or lumbar (lumbodorsal) muscles. Most are given in the caudal muscles of the thigh, the semimembranosus and semitendinosus (Figure 3.25). Movement during the injection should be minimized by the person giving the injection firmly holding the front of the thigh.

ADMINISTRATION OF ORAL, OPHTHALMIC, AND OTIC MEDICATIONS

- **Oral:** If it is safe and effective, hiding medication in food is always preferable to injections.
  - **Pills and Capsules**
    - **Hidden in Food Treats:** Most tablets can be crushed or capsules taken apart and the medication mixed in a pasty treat such as peanut butter, soft cheese, meat-based baby foods, cream cheese, hot dogs, or tuna paste.

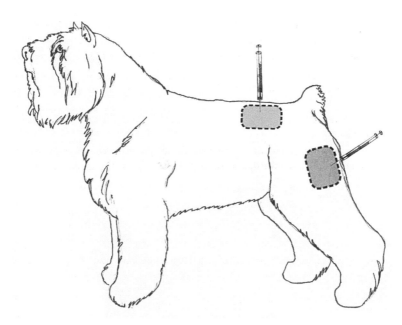

Figure 3.25   Most common sites for IM injections in dogs.

- **Preparatory Training:** Dogs should be trained for pilling by occasionally using small treats in which a tablet, capsule, powder, or liquid may be hidden in the future. A small portion of the food vehicle should be offered prior to the medicated portion and followed after the medicated portion by another small portion of the unmedicated food vehicle.
- **Manual Administration** (*Procedural Steps 3.27*)

Procedural Steps 3.27	Manual Administration of Oral Medications to Dogs
1.	Oral administration of medications should be done with the dog in sitting position.
2.	Large dogs should be straddled facing out of a corner of a room to prevent them from being able to back up.
3.	If holding moderate-sized dogs, the handler straddles and kneels behind them.
4.	The handler's lower abdomen or thigh blocks their attempts to back up.
5.	Small dogs can be held similarly on an exam table by the handler standing at the end of the table and having the dog sit with its back against the handler's abdomen.
6.	After restraining the dog's body, the handler grasps its head and upper jaw with his left hand and presses the cheeks between the teeth while pulling down on the lower incisors with the index finger of the right hand.
7.	The thumb and middle finger of the right hand hold the tablet or capsule.
8.	When the dog's mouth is open, the handler places the medication on the back part of the tongue and closes the mouth (Figure 3.26).

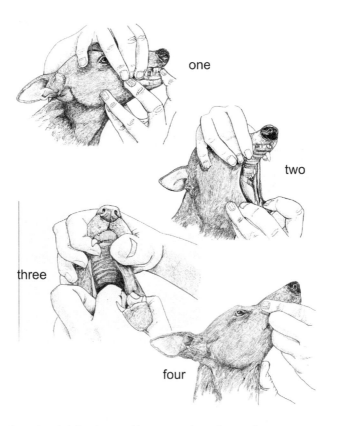

one

two

three

four

Figure 3.26   Steps in administering a tablet or capsule orally to a dog.

9.	Swallowing can be encouraged by holding the nose up while stroking the throat or blowing on the nose.
10.	The dog should be observed for signs of swallowing or licking its nose, which also suggests swallowing the medication.
11.	A liquid (water, broth, tuna juice) should be administered afterward as counterconditioning and to reduce the risk of the medication lodging in the esophagus.

- **Pill Gun:** A pet *pill gun* or syringe is similar to a syringe with a rubber end that entraps the pill and a plunger to dislodge the tablet or capsule after it is placed over the hump of the tongue when the mouth is open. Pill guns should be thoroughly washed and allowed to completely dry after each use.
- **Liquids:** Oral administration of liquids should be done with an oral syringe, not a cup (***Procedural Steps 3.28***).

Procedural Steps 3.28 Oral Administration of Liquids to Dogs	
1.	The dog is restrained in the same manner as for tablets or capsules but with its head held up with the mouth closed.
2.	The syringe is placed in the dog's cheek pouch, and the medication is injected slowly enough to permit the liquid to run back past the teeth and for the dog to swallow the liquid without difficulty (Figure 3.27).
3.	Liquids should never be injected over a dog's tongue with its mouth open due to risk of it entering the larynx. Irritation of the larynx or aspiration pneumonia may result.
4.	If small volumes are being administered, feeding a slice of bread with peanut butter on one side and the liquid medication soaked into the other side of the slice may suffice.

- **Ophthalmic**
    - **Risks:** Poor restraint can allow a dog to suddenly shake its head and accidently be poked in the eye by a handler attempting ophthalmic administration.

Figure 3.27   Administering a liquid into the cheek pouch.

- **Preferable Method:** To medicate the eye, a handler must usually restrain the dog in sitting position and hold its head and front legs while the person administering the medication holds the upper and lower eyelids apart.
- **Alternative Method:** A less desirable method requires just one person. The handler stands behind the dog and grasps its muzzle with one hand and places the heel of his other hand on top of the dog's head while holding the medication to be applied to the face or eyes. A finger of the hand on the muzzle pulls a lower lid down, and the thumb of the hand on the top of the dog's head can pull an upper lid up.

- **Otic:** To medicate an external ear canal, a handler restrains the dog in a sitting position while the person administering the medication holds the ear flap (pinna) upward and outward with the heel of the hand on top of the dog's head. Docile dogs with nonpainful ears will allow examination and treatment by one person with little restraint required.

CYSTOCENTESIS

- **Definition:** Cystocentesis is collection of urine using a needle and syringe to collect the urine by puncturing the abdomen, entering the urinary bladder, and aspirating the urine.
- **Restraint Position:** Restraint positions should be ventrodorsal or upright standing. These positions stretch the body and limit torso movement. Standing in horizontal, quadruped position or lateral recumbency permits too much movement and possible internal slashing injury if struggling occurs during the procedure.

SPECIAL EQUIPMENT

HEAD COLLARS

Head collars (halter-type collars) are used for training dogs not to pull on a leash.

- **Construction:** Head collars have straps that go behind the head and over the nose just in front of the eyes; the leash attaches to a ring below the throat (Figure 3.28).

Figure 3.28   Head collar.

- **Action**
  - **Attention to the Handler:** Pulling on the leash applies pressure to the back of the head, pulls the nose down, turns the head, and closes the mouth. The effect is based on a dominant dog's method of grasping the muzzle of a subordinate dog to establish or reassert its higher social rank. The movement insists on attention to the handler.
  - **Pulling by the Dog Is Less Effective:** The pulling power of the muzzle is much less than the pulling power of a collar on the neck.
  - **Detailed Fitting Required:** To be effective, more detailed fitting is required than for other collars.
- **Use:** To properly use a head collar, tension on the leash should be steady. Jerking and release of the leash is not appropriate. These are not true restraint devices and should not be left on when not attached to a hand-held leash.

BEHAVIOR JACKETS

- **Compression Jacket:** Compression on the trunk of the body by a snug jacket has been claimed to be 80% effective in controlling anxiety in dogs, particularly from thunderstorms. However, adaptation or desensitization occurs with a loss of effect in about 20 minutes. Compression jackets should never be used in dogs with respiratory problems.
- **Static Control Jacket:** Jackets with a metallic lining have been promoted to discharge static electricity in the haircoat, causing improvement in anxiety related to static electricity. Independent, controlled studies proving efficacy have not been reported.

CAPTURE POLES

Capture poles (inappropriately called *rabies poles*) are hollow metal handles with a rope or wire cable that is fixed at the catching end, which forms a loop and goes through the handle to the handler's end (Figure 3.29).

Figure 3.29   Use of a capture pole stabilized by pushing the end into a corner.

- **Risks to the Dog**
  - **Compression of the Trachea:** The loop is used to catch dangerous dogs by the neck and can injure the trachea (windpipe) if used with too much force.
  - **Neck Fractures:** Dogs (or any other animal) should never be lifted by the neck alone using a capture pole, because severe damage may occur to the trachea or because struggling when dangling could break its neck.
  - **Wire Cuts and Punctures:** The loop is a coated cable, which should be checked each time before use to ensure no wire strands have become frayed by prior use and become exposed, which could injure the animal's neck.
- **Risks to the Handler: Whip Lashes**
  - **Cable Whip:** The loop is spring released by a knob on the handler's end. Uncontrolled rapid recoil can allow the cable to whip around, endangering the handler or bystanders.
  - **Prevention:** The handler should hold the release knob with his first two fingers and the wire with the last two fingers and heel of his hand. This permits control of the wire, preventing it from whipping when the loop is released.
- **Method of Use (*Procedural Steps 3.29*)**

Procedural Steps 3.29	Proper Use of Capture Poles
1.	To apply a capture-pole loop, the handler must restrict the evasive movement of the dog as much as possible.
2.	It is important to have a slow, deliberate, and persistent approach with the pole.
3.	The dog's head should be approached slowly with the loop to reduce the risk of bumping its head, eyes, or teeth with the end of the pole.
4.	Dogs often will bite at the pole, which should not be withdrawn when this happens due to the possibility of injuring the dog's teeth and seeming submissive to the dog's actions.
5.	If the pole is not withdrawn when the dog bites, it will eventually ignore it.
6.	Some dogs learn to lower their head to the floor as an evasive maneuver, so the handler must be patient and slowly continue to get the loop to slip over the nose and then the head.
7.	Once the dog has been captured, the dog's end of the pole should be pushed into a corner to control its body movements, since otherwise, a strong dog has considerable leverage at the end of the pole and ability to move.
8.	Dogs should not be led or dragged by a capture pole.

LEATHER GLOVES WITH GAUNTLETS

- **Bite Protection:** Gloves with forearm covers (gauntlets) can sometimes be useful in handling small dogs that bite. Gloves should be large enough that the handler's fingers do not extend to the end. The extra length likely to be bitten by a dog can reduce danger to the fingers.
- **Disadvantages:**
  - **Excessive or Inadequate Restraint Pressure:** Care must be taken not to restrain a small dog too tightly while wearing gloves. Gloves can reduce the sensitivity of the handler's hands to the amount of pressure being applied.
  - **Retained Odors and Poor Sanitation:** Restraint gloves are difficult to clean, and the odors from previously restrained dogs on gloves and gauntlets may increase the stress and apprehension of some dogs.

HEAD MOVEMENT–LIMITING DEVICES

- **Elizabethan Collars:**
  - **Description:** Elizabethan collars are named for the large collars on dresses that were made fashionable by Queen Elizabeth I in the 1550s. Elizabethan collars

for dogs are cone-shaped collars that fit around the neck with the outer edge toward the dog's nose. The head is surrounded 360 degrees on the sides, top, and bottom by the collar, preventing the dog from chewing most of its body, although it may still reach its front feet.

- **Disadvantages:**
  - **Impaired Sight and Hearing:** Peripheral sight and hearing are impaired by the collars.
  - **Impaired Ability to Eat or Drink:** Some dogs cannot reach food and water with Elizabethan collars, and the collar must be removed often to allow eating and drinking.
  - **Obstructive to Movements:** Collars are bulky and will catch on doors, furniture, scrubs, and other objects.
  - **Potential to Injure Neck:** Poorly made or fitted collars can injure the neck with sharp edges. Any rough or sharp edge should be well padded with layers of medical adhesive tape.

- **Globe Collars:** Spherical semiopaque (globe-shield) collars that encapsulate the head and have an opening in front for breathing and vision are commercially available as a means to protect handlers from bites. Spherical collars do not allow drinking or eating and must be frequently taken on and off, which could increase the risk of being bitten.

- **Broad Neck Collars:**
  - **Description:** Thick, broad collars can wrap snugly around the neck like a human cervical collar to limit the dog's ability to reach areas of its body. Some are pneumatic (Figure 3.30). A retention strap is run from the top of the broad band, around the lower aspect of the thorax, and behind the front legs (Figure 3.31).

Figure 3.30 Pneumatic movement-limiting collar.

Figure 3.31   Broad-band movement-limiting collar.

- **Advantages and Disadvantages:** Broad-band neck collars permit better peripheral vision and hearing and do not catch on objects as does the Elizabethan collar. Eating and drinking are not blocked. However, access to most of the front legs is possible.

SHOCK COLLARS

- **Description:** Shock collars are electronic collars with metal contact points that press on the skin of the neck. A remote control operated by the handler activates a shock when desired. Shock collars are often mistakenly viewed by inexperienced handlers as a shortcut to training or a quick fix for bad behavior.
- **Disadvantages:**
  - **Can Cause Aversive Behavior:** Shock collars often cause aversive behaviors in dogs.
    - **Shock Must Be Introduced:** Vibration warnings before a shock and adjusting the voltage before use are proposed safety features. However, warning vibrations are ineffective if an association with a following shock is not first established.
    - **Voltage Is Not Instantly Adjustable:** If a shock is delivered, the voltage is not quickly adjustable to the situation as is use of a slip-chain collar. Dogs that get wet during training can get a higher-than-intended voltage.
    - **Detrimental to Dog Socialization:** For the above and other reasons, shock collars should never be used on puppies.
    - **Illegality:** Shock collars are prohibited in Germany, Denmark, Norway, Scotland, Sweden, Switzerland, Wales, and Canada (Quebec).

- **Last-Option Salvage:** The use of shock collars should be limited to experienced trainers with impeccable timing as a tool to consider if traditional methods are not effective in improving a seemingly incorrigible adult dog's bad behavior. In selected cases, proper short-term use of a shock collar by an experienced trainer may salvage a dog from being relinquished to a shelter or euthanized.

VIBRATING, SPRAYING, AND ULTRASONIC COLLARS

- **Vibrating Collars:** Vibrating collars work with a remote control similar to shock collars without the risk of pain to the dog and aversion. They are not reliable training tools, since many dogs can become desensitized to vibration and ignore the stimulus. However, they can be helpful in providing command signals to a deaf dog.
- **Spray or Ultrasonic Collars:** Collars that emit a spray of citronella or an ultrasonic noise are also available. Typically, these are used to discourage barking. They are activated when any dog within sensor range barks and can therefore deliver a spray or ultrasonic burst to a nearby innocent dog.

BLINDFOLD CAPS

- **Theoretical Basis:** Blindfolds of sheer fabric attached to a dog's head and eyes like a cap have been suggested to have a calming effect on dogs. Blindfolds are effective in horses, but dog caps for blindfolding dogs are not opaque, and dogs do not depend as much on their sight for assessing potential threats as do horses.
- **Efficacy:** Evidence of the effectiveness of dog blindfold caps is only anecdotal, but dimming light with a transparent blindfold (or otherwise) may have calming effects on some dogs in a quiet environment with no strange odors present. However, impeding a dog's vision by blindfolding of any type may also exacerbate startle responses.

MOBILITY ASSISTANCE

- **Description:** Dogs with an inability to control their hind legs (paraparesis) or move their hind legs (paraplegia) require assistance in walking.
- **Forms of Assistance:** Assistance in ambulation can be provided with a rear or body harness or a support sling. In the absence of either, a towel or belt can be used as a support sling under the abdomen. Two-wheel carts that support the dog's midsection and hindquarters permit mobility even when the dog cannot otherwise stand on its hind legs.

TRANSPORTING DOGS BY AUTOMOBILE OR TRUCK

Dogs are transported more than any other pet animal. Most transportation of dogs is by car or truck. Some dogs are transported by airlines. Amtrak only allows service dogs to travel.

REGULATIONS

Regulations on interstate, international, and air travel change frequently and must be rechecked each time a dog is transported. Import, export, and interstate transportation information on privately owned dogs is available at www.aphis.usda.gov/aphis/pet-travel.

PREPARATION FOR TRAVEL

- **Possible Escape**: Regardless of the means of travel, there is risk of a medical emergency or escape.
  - **Have One or More Means of ID**
    - **Harness and ID Tag:** A chest harness should be worn with an ID tag that includes the owner's name, address, and cell phone number. A landline

phone number, while traveling, is much less useful. A chest harness is more secure and prevents escapes better than a collar.
- **Tattoo:** Dogs that travel should be tattooed with the owner's cell phone number in their ear flap or inner surface of a hind leg. Tattoos may fade and need to be refreshed before travel to be easily read.
- **Microchip:** Unlike a tattoo, an embedded microchip is not immediately able to be read by someone who may try to rescue an escaped dog, but a microchip is a good second choice to ensure an escaped dog's return. Microchips are required in dogs in Japan, the European Union, and many other nations.

- **Pictures:** Current pictures of the dog should be kept ready to distribute to searchers in case a search becomes necessary for the dog after an escape.
- **Provide Destination Information:** A travel tag should also be attached to the harness or collar that provides the destination and destination contact information.
- **Have Accessible Food, Water, and Medications:** Food, water, and any needed medications should also be available during the trip.

TRAVEL BY CAR OR TRUCK CAB

- **Prepare in Advance:** Dogs should be desensitized to car travel by experiencing frequent short trips by car to a pleasurable destination with no adverse events during their primary socialization period (6 to 12 weeks of age) or as soon after that as possible. If they will be transported in a carrier, pretrip training should include free access in and out of the carrier, being fed in the carrier, and sleeping in the carrier to develop a feeling of security while in the carrier.
- **Immobilize Outside Range of Air Bags:** If transporting by car or truck cabs, dogs should ride in a back seat and be restrained in a restraint harness fastened to a seat belt buckle or in a crate that is strapped to floor anchors. This protects dogs from airbag injuries and the driver from interference with driving (vision obstruction, interference with braking) or being injured by a dog becoming a missile in an accident. More information on travel restraints is available at the Center for Pet Safety: www.centerforpetsafety.org/.
- **Allow Rest Stops:** The handler should stop every 2 to 3 hours for the dog to exercise and eliminate. If the weather is hot, water should be provided in the shade in plastic water bowls that cannot tip over.
- **Avoid Temperature Extremes:** If the temperature is over 72°F, dogs should not be left in cars. At 72°F, the inside of a car can reach 100°F in 30 minutes. Temperatures below 55°F may be too low for some dogs. Avoid clothing dogs with sweaters or coats due to the risk of overheating.
- **Keep Windows Up:** A dog should not be allowed to ride in a car with its head out of a window. It may become excited and jump, which is hazardous to the dog and to other traffic. Eye injury from flying insects or other flying objects are common in dogs that stick their head out car windows.
- **Health Regulations for Interstate Travel** (Table 3.16)

Table 3.16 U.S. Federal Requirements for Interstate Travel of Dogs	
•	Certificate of veterinary inspection
•	Provision of adequate shelter from all elements and protection from injury
•	Sufficient cleanliness to avoid contact with urine and feces
•	Protection against hazardous temperature extremes
•	Uncontaminated and nutritious food at least once per day
•	A program of parasite control

TRAVEL IN PICKUP TRUCK BED, CAMPERS, AND TRAILERS

- **Improper Transport:** Dogs should not be loose or tethered in pickup beds due to the risks of being thrown out; injured by sliding around or by shifting cargo; getting eye, ear, or mouth injuries from wind and debris; and, if tethered, choking. An estimated 100,000 dogs die per year in the U.S. by jumping or falling from truck beds. Burns may occur from sun-heated metal.
- **Proper Transport:** Dogs can be safely transported in commercial kennels (also called boxes) for pickup trucks that are properly mounted, shielded, insulated, and ventilated.

NOTE

Additional recommended readings on dog handling are available in references on multiple species of small animals provided in the Appendix.

DOG HANDLING REFERENCES AND SUGGESTED READING

1. American Veterinary Medical Association. Model Bill and Regulations to Assure Appropriate Care for Dogs Intended for Use as Pets. www.avma.org

2. Carter A, McNally D, Roshier A. Canine collars: An investigation of collar type and the forces applied to a simulated neck model. Vet Rec 2020;187. doi:10.1136/vr.105681

3. Case L. Perspectives on domestication: The history of our relationship with man's best friend. J Anim Sci 2008;86:3245–3251.

4. Cooper JJ, Cracknell N, Hardiman J, et al. The welfare consequences and efficacy of training pet dogs with remote electronic training collars in comparison to reward based training. PLoS ONE 2014;9:e102722: doi:101371/journal.pone.0102722

5. Cutler JH, Coe JB, Niel L. Puppy socialization practices of a sample of dog owners from across Canada and the United States. J Am Vet Med Assoc 2017;251:1415–1423.

6. De Keuster T, Lamoureax J, Kahn A. Epidemiology of dog bites: A Belgian experience of canine behaviour and public health concerns. Vet J 2006;172:482–487.

7. Drobatz K, Smith G. Evaluation of risk factors for bite wounds inflicted on caregivers by dogs and cats in a veterinary teaching hospital. J Am Vet Med Assoc 2003;223:312–316.

8. Frank D, Lecomte S, Beauchamp G. Behavioral evaluation of 65 aggressive dogs following a reported bite event. Can Vet J 2021;62:491–496.

9. Kinsman RH, Casey RA, Knowles TG, et al. Puppy acquisition: factors associated with acquiring a puppy under eight weeks of age and without viewing the mother. Vet Rec 2020 Aug 8;187(3):112. doi:10.1136/vr.105789. PMID: 32764003; PMCID: PMC7456714.

10. McMillan FD, Serpell JA, Duffy DL, et al. Differences in behavioral characteristics between dogs obtained as puppies from pet stores and those obtained from noncommercial breeders. J Am Vet Med Assoc 2013;242:1359–1363.

11. Patronek GJ, Slavinski SA. Animal bites. J Am Vet Med Assoc 2009;234:336–345.

12. Patronek GJ, Sacks JJ, Delise KM, et al. Co-occurrence of potentially preventable factors in 256 bite-related fatalities in the United States (2000–2009). J Am Vet Med Assoc 2013;243:1726–1736. (Response in Letters to Editor, Raghavan M, et al. J Am Vet Med Assoc 2014;245:484.)

13. Raghaven M, Martens PJ, Chateau D, et al. Effectiveness of breed-specific legislation in decreasing the incidence of dog-bite injury hospitalizations in people in the Canadian province of Manitoba. Inj Prev 2012:19:177–183.

14. Sacks JJ, Sinclair L, Gilchrist J, et al. Breeds of dogs involved in fatal human attacks in the United States between 1979 and 1998. J Am Vet Med Assoc 2000;217:836–840.

15. Shuler CM, DeBess EE, Lapidus JA, et al. Canine and human factors related to dog bite injuries. J Am Vet Med Assoc 2008;232:542–546.

16. Stellato AC, Dewey CE, Widowski TM, et al. Evaluation of associations between owner presence and indicators of fear in dogs during routine veterinary examinations. J Am Vet Med Assoc 2020;257:1031–1040.

17. Yin S. Low Stress Handling, Restraint and Behavior Modification of Dogs & Cats. CattleDog Publishing, Davis, CA. 2009.

18. Yin S. Simple handling techniques for dogs. Compend Contin Educ Vet 2007;29:352–358.

4

CATS

DOI: 10.1201/9781003110927-4

The domestic cat is the most numerous companion pet in the U.S., although fewer households have cats than dogs. Male cats are called **tomcats**. Female cats are **queens**, and young cats are **kittens**.

NATURAL BEHAVIOR OF CATS

Feral and domesticated cats are highly social, nocturnal, territorial, semiarboreal, solitary predators. They prefer to sleep 16 to 18 hours a day. Females are more social than males.

BODY LANGUAGE

- **Relaxed** (Table 4.1):

Table 4.1 Body Language of a Friendly Cat
• Lying on one side or sitting while its tail moves slowly
• The tail hangs down in a relaxed manner when casually walking
• Kneading of soft surfaces
• If greeting an unthreatening animal or human, the tail is carried up
• Rubbing (bunting) against humans or other animals
• Purring

- **Agitation:** A cat on alert is characterized by assuming a frozen sitting or lying posture, rapid flicking of the tail, and dilated pupils. Walking tiptoe with head down is aggressive posture. Aggressive body language also includes slight piloerection on back and ears erect but swiveled to the side or back against the cat's neck.
- **Fear:** Fear is exemplified by attempts to hide, flattened ears, crouching, arching the back, salivation, spitting, and dilated pupils.
- **Lure for Aggressive Play:**
  - **Exposing Abdomen:** Cats will play fight to determine social rank. A common technique is to lie on a side with their abdomen exposed and head up. An approach may trigger a scratch or bite.
  - **Caution for Handlers:** This type of aggression may be directed toward a handler when he tries to pet a cat's abdomen. Efforts to pet a strange cat's abdomen should be avoided even if the cat seems to be inviting it.

VOCALIZATIONS

Purring is a sign of contentment. Cats will often chatter their teeth, also called a chirrup, if they are excited by the sight of prey. Meows are to call the attention of other cats or their handler.

MARKING TERRITORY

- **Scratching:** Cats instinctively scratch objects in their territory to mark their territory, stretch their muscles, and clean and sharpen their nails. Soft wood is preferred. Scratch marks are a visual marker, but pheromones from the cat's paws also provide an olfactory marker.
- **Urine:** Both males and females will mark territory by spraying urine. Males spray to mark territory; females spray while in heat to attract males.
  - **Triggers:** Urine spraying of territorial objects is intensified with the introduction of a rival cat, and territorial aggression is often triggered by another cat's odor.
  - **Urine Odor on Handling Jackets:** Clean (deodorized) cat handling jackets should be worn when handling an aggressive cat to avoid a territorial aggression response.
- **Scent:** Cats rub with their cheeks (called *bunting)* when objects stimulate a gape or when a subordinate greets a dominant cat. Scent glands next to their mouths produce chemicals that are smeared on objects (and handlers) that the cat claims as its own by facial marking.

HANDLER SAFETY

- **Common Cat Behavior:**
  - **Usually Friendly:** Most domestic cats are inherently friendly. A few cats are always ill-tempered.
  - **Agile and Quick:** If they are in good health, all are agile, extremely quick, and capable of causing serious injuries to handlers. Minimal restraint for the procedure to be done is the best means of handling cats.
- **Aggression toward Owner:** The most frequent reported behavior problem of cats is aggression toward its owner. Most often, the cause is poor socialization of the cat as a kitten and poor handling techniques of owners, in particular, excessive restraint and rough play.
- **Dangers for Handlers from Cats:** The dangers from handling cats are bites, which are more common in veterinary personnel than dog bites, plus scratches and infections. At least 28% of small animal veterinarians and veterinary technicians are bacteremic with *Bartonella* species, the Cat Scratch Disease bacteria. Ringworm is the most common zoonotic disease acquired from cats.
  - **Bites:** Cat bites are more common in veterinarians and veterinary technicians than dog bites; although they do not pose a risk to human life, they can inflict serious injuries that may lead to impaired use of the hands.
  - **Claws:** Cats' first line of defense is their front claws, which can cause painful injuries and impairments to a handler's hands, arms, and eyes.
  - **Infections:** Even superficial cat scratches can introduce bacteria, such as *Bartonella*, the bacteria for Cat Scratch Disease, or a subcutaneous fungus called *Sporotrichum*.
- **Handler Attire:** Lab or clinic coats with long sleeves should be worn when handling cats as a means of protecting against cat scratches. Back claws are a source of injury to handlers when holding a cat near the handler's body if the cat attempts to escape.
- **Risks to Hands:**
  - **Speed of Bites:** Cats bite very quickly and let go quickly. They then will bite quickly again if the threat does not retreat.
  - **Depth of Bites:** The bites are deep penetrating wounds that can injure and infect joint capsules, tendon sheaths, and bones, particularly of the hands. Permanent disabilities of the hand can result from cat bites.
  - **Do Not Simulate Hands as Prey:** Socialization with humans involves handling and playing with cats, but play should not involve using hands as simulated prey. *Fishing* play with cats using a rod, string, and feathered object is much safer.
- **Fear Retreat:** Domestic cats are very independent, especially if threatened.
  - **Aimless Fleeing:** Their first reaction to a threat is to run and hide with no regard to where other cats are running or to other potentially dangerous things going on in the same area. In other words, cats run first and think later.
  - **Defensive While Hiding:** Once hidden as well as they can, they often will issue warnings (low rumbling growls, hisses, and rapid strike and retreat) to threats that continue to approach.
- **Signs of Agitation:**
  - **Sensitivity of Detection:** Cats telegraph their aggression more consistently than do dogs.
  - **Typical Signs:** A dominance aggressive cat may do little cowering or hissing before striking, but they will have a fixed stare toward their opponent, dilated

eyes, and their ears will be pulled back. They will stand confidently. Their tail will move back and forth to the sides with a flicking movement at the end of the tail. Their hair on the back will be raised, and their whiskers are elevated to a position where they stick straight out to the sides.

- **Fear Aggression:** Fearful aggressive cats are more vocal and will flatten their ears and arch their back before striking, usually from a crouching position. They do not stare directly at the opponent and may present their side to what they perceive as danger.

CAT SAFETY

- **Socialization**
  - **Socialization to Humans:** Cats that are socialized to people have less resistance to being handled and restrained. As a result, they are safer from self-inflicted or inadvertent injury from attempted escapes when being handled.
  - **Period of Critical Socialization:**
    - **Preparation for Socialization:** Kittens should begin their routine vaccinations 10 days prior to beginning the first socialization event with other cats, and they should be tested negative for feline leukemia virus (FeLV) and feline immunodeficiency virus (FIV). If adopted from a shelter, the kitten should be kept in their new home for 2 weeks before socializing with other cats. Kittens should not socialize with other cats that are sneezing, coughing, vomiting, or having diarrhea.
    - **Socialization with Humans:** Kittens should be socialized to other animals and humans outside the immediate family during the sensitive period for cats of 2 to 7 weeks of age.
    - **Socialization with Other Cats:** Intercat socialization is particularly important during 9 to 14 weeks of age. It is during this period that their focus shifts from social play to predatory hunting practice.
    - **Social Maturity:** Adult social maturity occurs by 36 to 48 months of age.
- **Prevent Outdoor Roaming:** Cats should be confined indoors or in appropriate outdoor containments, such as catios/screened porches or mesh exercise tunnels (Table 4.2)

KEY ZOONOSES

Apparently healthy domestic cats pose little risk of transmitting disease to healthy adult handlers who practice conventional personal hygiene.

- **Physical Injury More Common:** The risks of physical injury are greater than the risks of acquiring an infectious disease. However, there is much overlap. Up to 80% of cat bites become infected. Some diseases can be transmitted by apparently healthy cats (Table 4.3)
- **Sick Cats:** Suspected sick cats may require special handling. Proper personal protective equipment for the illness should be determined by the veterinarian in charge of the case and communicated to all handlers.

Table 4.2 Disadvantages of Cats Roaming Free Outdoors
• Exposure to infectious diseases, including rabies, feline leukemia virus, feline immunodeficiency virus, feline rhinotracheitis virus, etc.
• External and internal parasites
• Being stolen
• Accidental or malicious injury from cars or trucks, other cats, dogs, and wildlife, including coyotes, foxes, alligators, cougars, and large raptors

Table 4.3 Diseases Transmitted from Healthy-Appearing Cats to Healthy Adult Humans

Disease	Agent	Means of Transmission	Signs and Symptoms in Humans	Frequency in Animals	Risk Group*
Bites and Scratches	—	Direct injury	Bite and scratch wounds to hands and arms. Infection from cat bites, especially with *Pasteurella multocida*, is much higher from cat bites than from dog bites.	All cats are capable of inflicting bite wounds and, if not declawed, scratch injuries.	3
Cat Scratch Disease	*Bartonella henselae*	Direct by scratches	Usually mild transient illness with enlarged regional lymph nodes	Occurrence is high in young cats.	2
Ringworm	*Microsporum canis*	Direct and direct from fomites	Ring of skin inflammation with red border	Most cats carry *M. canis*.	2
Visceral larvae migrans	*Toxocara cati*	Direct from soil, fecal-oral	Abdominal pain—inflamed liver	Most kittens will have *T. cati*.	2
Toxoplasmosis	*Toxoplasma gondii*	Direct from soil, water, or litter, fecal-oral	Usually mild transient illness in an adult. Concern is greatest for pregnant women and the fetus.	Approximately 30% of cats allowed outdoors have carried toxoplasmosis.	2
Echinococcus	*Echinococcus multilocularis*	Direct, fecal-oral	Lung cysts that cause shortness of breath and coughing	Can occur in cats allowed to eat wild rodents in northern states of the U.S.	4

* Risk Groups (National Institutes of Health and World Health Organization criteria. Centers for Disease Control and Prevention, Biosafety in Microbiological and Biomedical Laboratories, 5th edition, 2009.)

1: Agent not associated with disease in healthy adult humans.

2: Agent rarely causes serious disease, and preventions or therapy possible.

3: Agent can cause serious or lethal disease, and preventions or therapy possible.

4: Agent can cause serious or lethal disease, and preventions or therapy are not usually available.

SANITARY PRACTICES

- **Zoonosis Prevention**
  - **Do Not Allow Roaming Outdoors:** Cats are instinctive predators. If allowed to roam outdoors, not only are songbirds at risk, but cats can acquire toxoplasmosis and external parasites, among many other zoonoses.
  - **Dress, Vaccinations, Control of Parasites:**
    - **Proper Attire:** A handler of cats should wear appropriate dress to protect against skin contamination with hair and skin scales or saliva, urine, and other body secretions. Gloves should be worn when handling cats if there are any open wounds on the handler's arms or hands.
    - **Avoid Cat Saliva:** A cat should not be allowed to lick any open wounds on handlers.
    - **Control External Parasites:** External parasites, fleas, and ticks should be controlled.
    - **Keep Vaccinations Current:** Vaccinations in cats should be kept current against rabies. Cat handlers should be vaccinated against tetanus every 10 years.
  - **Practice Basic Sanitation**
    - **Clean Hands:** Basic sanitary practices should be followed, such as keeping hands away from eyes, nose, and mouth when handling cats and washing hands after handling them.
    - **Avoid Fecal Contamination:** Feces should be removed from litter boxes and disposed of daily.
    - **Clean Food and Water Bowls:** Cat food bowls and food utensils should be washed after each use.
- **Preventing Spread of Disease among Cats**
  - **Initial Control Measures:** When handling more than one cat from different households or catteries, proper sanitation is required to prevent the spread of disease from carriers without clinical signs.
    - **Isolate New Cats:** Cats from different origins should not be confined in the same cage or group area.
    - **Maintain Sanitation:** Basic procedures for handlers to wash their hands and to clean and disinfect tabletops and cages used in handling should be followed.
    - **Sanitize Potential Fomites:** Restraint equipment such as blankets, muzzles, capture poles, grooming equipment, collars, cat restraint bags, and slip leashes should be disposable or cleaned and disinfected. Leather gloves should be kept as clean as possible and used infrequently.
  - **Sick Cats:** Special precautions are needed if sick cats are handled, and sick cats should be isolated from apparently normal cats. New household or cattery members should be quarantined for at least 2 weeks to reduce the risk of transmitting a disease that new animals could be incubating before introducing to the rest of the clowder.

APPROACHING AND CATCHING

The cat's attitude should be observed before attempting to capture it. Most cats can be classified as nonaggressive or fear-aggressive. Most nonaggressive cats are still resentful of restraint and respond best to an unhurried approach and loose, gentle restraint.

NONAGGRESSIVE CATS

- **Handler Approach:** The handler should move slowly but with confidence, use a calm, assuring voice, and lower his body near the cat with his side toward it. Small bits of food treats can be used if needed to lure the cat closer to the handler. Cats should not be stared at or leaned over.
- **Cat Approach:**
  - **Direct Approach:** A friendly approach by a cat is with its tail held up with back legs slightly extended. Purring may be audible. It will almost touch the person it is approaching with its nose and may rub its face and head on the handler.
  - **Exposing the Lower Abdomen:** Rolling over and exposing its belly is not a submission sign in cats. It is an invitation to play, but touching the abdomen may trigger a playful bite.
- **Method (*Procedural Steps 4.1*):**

Procedural Steps 4.1	Method of Approaching and Catching Nonaggressive Cats
1.	An apparently friendly cat should be allowed to approach the handler's extended index finger, which mimics another cat's nose, to smell it.
2.	The handler can then quietly and slowly move his hand to stroke the cat's head.
3.	Stroking a friendly cat's back results in arching of their back to press more firmly against the stroking hand, a signal of invitation for more petting.
4.	A slip leash should be applied, and then the cat is moved so that it is in front of the handler, facing his right side.
5.	The handler's left hand reaches over the cat's back and grasps the cat's right front leg with the wrist supporting the cat's body to prevent escape from the handler's hold or the cat climbing up the handler's chest.
6.	The right hand holds the slip leash and is kept near the cat's head to pet it as it is carried and grasp the scruff of the neck if struggling occurs.
7.	The handler's right hand is used to comfort and distract the cat while being carried.

- **Distraction Techniques:** Distraction techniques can help minimize the needed restraint for handling cats (Table 4.4).

AGGRESSIVE CATS

- **Body Language:** The body language of an aggressive cat is that of piloerection, arched back, tail down with its tip flicked slowly, and ears erect and pointed forward. If approached, it will flatten its ears back, bat with paws, and lean away from the threat, vocalizing.
- **Two Handlers Needed:** Aggressive cats must be handled by two people. The primary handler has to concentrate on only the restraint. The other person performs the examination, administration of medications, or other procedures needed.

Table 4.4	Distraction Techniques for Handling Cats with Minimal Restraint
•	Gently rubbing the cat's head and ears
•	Scratching the ears or throat and chin
•	Gently and rhythmically tapping the cat's head or face
•	Blowing softly on the nose
•	Stroking or wiggling the cat's foot or leg

- **Surroundings Preparation:** The handling room should be prepared for possible escape attempts. All doors, windows, and cabinet doors must be closed. Access to vents, backs of refrigerators, chimneys, or any other escape or hiding area that will impede efforts to recapture the cat if it escapes during the first capture attempt must be blocked. Anything breakable or spillable on countertops should be removed.
- **Method** (*Procedural Steps 4.2*):

Procedural Steps 4.2	Method of Approaching and Catching Aggressive Cats
1.	If a cat is in defensive posture but does not attempt to strike and retreat, a loop from a slip leash should be dropped over the cat's head to provide a means of gently moving the cat toward the handler.
2.	The handler can then either stroke and pick the cat up or, if necessary, use additional capture means (wrap in a towel, pull into a transport crate or box, or administer chemical restraint).
3.	Use of a thick towel to begin the stroking and gradual wrapping of its body from the neck back may be effective.
4.	Thick leather gloves with gauntlets to protect the wrists and forearms are an alternative but less desirable, approach.

- **Fractious Cats:** Fractious cats that will attack when capture is attempted in a cage should be entrapped by a capture pole, cat tongs, nets, or a cat loop on a flexible rod. Cats can wiggle, roll, and spin in a net; therefore, gloves or towels may need to be used to hold the cat down to administer medications or sedatives.
- **Feral Cats:** Feral cats may be caught in humane cage traps, which are commercially available, or netted and transferred to a squeeze cage for chemical restraint to be safely handled.

HANDLING FOR ROUTINE CARE AND MANAGEMENT

BASIC EQUIPMENT
Tractable cats require no equipment for handling, but slip leashes should be used on cats whenever they are outside a cage to aid in positioning the cat to be picked up and to increase security against an attempt to escape. If additional restraint is needed, particularly for cats that have not been declawed, towels and blankets are basic equipment.

- **Slip Leashes**
  - **Description:** A slip leash is a rope, cord, or flat woven strap with a metal ring honda or tied honda knot used for routine handling of cats. Flat-strap slip leashes should not be used due to their inability to maintain an open loop when being placed over the cat's head and neck. A slip leash serves as a sliding collar and lead rope in one piece.
  - **Cautions:** Slip leashes should not be used on cats with breathing problems. If an alternative does not exist, the loop should be placed around the neck with one front leg through it to prevent pressure on the trachea.
  - **Choking Risk:** Cats should never be tied and left unattended with a slip leash, because either escape or strangulation may result.
- **Towels and Blankets**: If additional restraint is needed, particularly for cats that have not been declawed, towels and blankets are basic equipment (*Procedural Steps 4.3*).

Procedural Steps 4.3	Towel or Blanket Restraints for Cats
Method 1	
1.	The first wrap should be around the neck and then the rest of the body is swaddled to restrict movement of the cat's limbs, euphemistically called making a *kitty burrito* (Figure 4.1).
2.	A leg can be withdrawn for venipuncture or the cat held on its back with the head extended for jugular venipuncture.
Method 2	
1.	While standing behind the cat, the handler drops the blanket over the entire cat, including its head.
2.	Then he quickly entraps the cat using both his forearms to sweep in and *taco shell* capture the cat, pressing the blanket edges under the cat's legs.
3.	Rear escape is blocked with the handler's torso, and forward escape is blocked by the towel over the cat's head.
4.	The wrap is then used to immediately swaddle the cat in a burrito-style wrap.

- **Tables and Table Covers**
  - **Tables:** Cats prefer elevated positions to rest, but stainless steel exam tables are not well tolerated. Time handling a cat on a table should be kept to a minimum and avoided when possible.
  - **Table Covers:**
    - **Assess the Value:** Covering the table with a pad or towel will provide traction and insulation, but allowing a struggling cat traction for its feet may be a disadvantage for the handler. Whether to cover a table should be determined on an individual basis depending on the actions of the cat on an uncovered table.
    - **Must Be Sanitary:** Table covers should be cleaned and sanitized after each use. Warming table towels has been recommended, but the added benefits to handling cats by using warmed towels has not been objectively assessed. However, some owners may value the anthropomorphic appearance of comfort of warm towels for their cat.

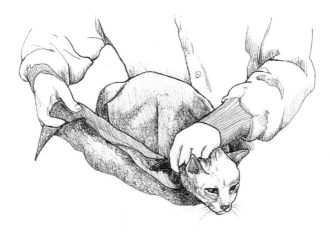

Figure 4.1   Towel wrap restraint.

MOVING

- **Carried in Arms:** Means of carrying cats in arms should prevent cats from escaping and climbing up the handler's chest (***Procedural Steps 4.4***).

Procedural Steps 4.4	Secure Procedures for Carrying Cats in a Handler's Arms
1.	When a handler carries a cat, the cat should have a slip leash applied first. It is lifted to gain control of the cats head (Figure 4.2).
2.	The body is then picked up with the handler's right palm under the cat's chest with an index finger between the front legs at the junction with the chest (Figure 4.3).
3.	The left hand, while holding the slip leash, is placed lightly on the back of the cat's neck and top of its shoulders.
4.	If distraction techniques (petting, scratching) are insufficient to control the cat, the cat can be scruffed with the left hand.
5.	An alternative hold is to allow the cat's sternum to be supported by the right wrist while the left foreleg is grasped with the right hand.

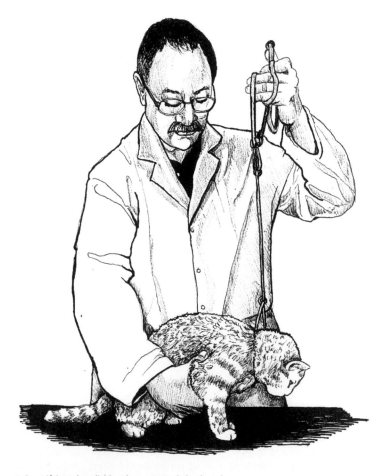

Figure 4.2   Lifting the slip leash to control the head.

Figure 4.3 Restraint for carrying a cat.

- **Transport in Crates**
  - **Transport Safety:** Transport crates are useful in securing cats when they are moved.
  - **Restricted Interaction:** Cats in crates in veterinary hospitals are protected from harm from other animals and from doing harm to other animals or humans. However, the cat will still be stressed by the near presence of other cats and dogs and should be kept from viewing them by covering the crate with a towel. Harsh lighting should be avoided, and the crate should be kept off the floor by placing it securely on a table or other elevated surface.

- **Construction and Contents:** Crates may be made of cardboard or plastic. Cat crates should have a top opening. Air holes should be present on at least 10% of the surface area of the crate. Absorbent bedding (towel or small blanket) should be provided.
- **Inserting the Cat:**
    - **Voluntary Entry:** Docile cats that have been previously acclimated to eating and resting in a crate should be given a chance to walk into a crate to seek seclusion and be offered a food treat. Putting a towel in may be an added lure, especially if the towel has been rubbed by the cat's owner, a buddy cat, or the cat to be crated to provide an odor of security.
    - **Method (*Procedural Steps 4.5*):**

Procedural Steps 4.5	Inserting a Cat into a Crate
1.	Failing a voluntary entrance to the crate, the cat should be picked up and placed in transport crates through a top opening, rump first (Figure 4.4).
2.	After the cat stands in the crate, the handler's hand should remain on the cat's neck and shoulders as the other hand closes the top against the handler's forearm.
3.	The restraining arm can then be slipped out of the crate and the top opening closed.
4.	If a top-loading crate is not available, the front of a front-opening crate can be tipped up, and the cat is lowered, rump first, into the crate.
5.	After closing the crate door, the handler should gently reposition the crate in a normal position.

Figure 4.4   Inserting a cat in a travel crate.

- **Removing the Cat** (*Procedural Steps 4.6*)

Procedural Steps 4.6	Removing a Cat from a Crate
1.	Removal of the cat should be done in a closed room with all exits and hiding areas blocked.
2.	The top lid should be gradually opened while gaining control of the cat with one hand. If the cat is tractable, it can then be lifted out using both hands (Figure 4.5).
3.	If the cat is agitated, it should be gently scruffed with one hand and its body supported with the other hand.
4.	If the cat is aggressive, a towel can be slid across the top door of the crate or between the top and bottom halves of a clamshell crate as a shield and visual barrier to the cat while the handler grasps the cat with the towel.
5.	Nets or gloves should be avoided if possible.
6.	The cat should be lifted out rather than allowing it to jump out on its own, and it should never be dumped out or scared out by thumping on the crate.
7.	If the crate does not have a top opening, the handler can turn the front door end up, or if it is a clamshell crate, the top half can be taken off of the carrier.

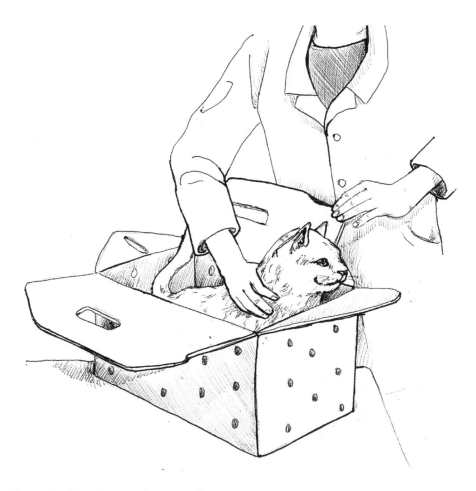

Figure 4.5   Removing a cat from a travel crate.

MANAGEMENT INVOLVING CAGES

- **Placement of Nonaggressive Cats in Cages**
  - Method (*Procedural Steps 4.7*):

Procedural Steps 4.7	Inserting a Cat into a Cage
1.	Cats should be placed in cages headfirst.
2.	The slip leash should be removed.
3.	One hand should have control of the cage door.
4.	Closure of the door should begin before release of the cat with the other hand so that when the restraint hand is removed, there is insufficient room for the cat to escape.

- **Release:** Release should be as smooth and quiet as possible, since this will be the most persistent memory of being last handled when the cat must be later removed from the cage. Struggling during the release will result in greater struggling when recapture is later needed. Control of the cat when placing it in the cage can be done with the restraint hand under its chest.

- **Removal of Nonaggressive Cats from Cages**
  - Method (*Procedural Steps 4.8*):

Procedural Steps 4.8	Removing a Cat from a Cage
1.	The handler should approach the cage in a friendly manner, speaking to the cat in a normal tone.
2.	Removal should begin with using one hand to open the cage door only enough to get the other hand and a slip leash in.
3.	Place the slip leash over the cat's head.
4.	After the cat's head is controlled, the cage door can be opened wider and the cat assisted by picking it up.

- **Removal:** Gently lift the cat's neck by raising the slip leash so there is direct control of the head before reaching under the cat's torso to lift it.

TRIMMING NAILS

Cat nails are sharp and curved for easy hooking penetration of the skin. They are dangerous in their ability to cause serious physical injury and potential for producing infected wounds. Cat Scratch Disease can be life-threatening in immunosuppressed people or children with immature immunity.

- **Purpose and Practicality:** Although controversial, trimming the nails every 2 to 4 weeks has been proposed by some as an alternative to onychectomy (declawing) by a veterinary surgeon experience in performing laser onychectomies. If successful, trimming nails of a housecat may reduce damage to furniture, but attempting to trim some cats can put the handler at greater risk of clawing injury or infection. Trimming is required more often in cats than in dogs because only a small portion of the nail can be safely trimmed in cats compared to dogs.
- **Method** (*Procedural Steps 4.9*):

Procedural Steps 4.9	Trimming Cat Nails
1.	If trimming nails is considered the best means to prevent scratch injuries, infections, and destruction of property, trimming can be attempted by one or two handlers.
2.	A cooperative cat can be held in the handler's lap with the handler's forearms blocking the cat's movements.

3.	One hand restrains a foot while placing the index finger on a claw's digital pad and then using the thumb to gently push the top of the nail to extrude it from the nail fold. The other hand uses a nail-trimming instrument.
4.	A second method is done with the cat on a table. The assistant handler holds the cat in a sitting or standing restraint while the primary handler holds a foot and extrudes and trims the nails.

HANDLING FOR COMMON MEDICAL PROCEDURES

Most handling and restraint of cats for medical procedures can and should be done without tranquilization, sedation, hypnosis, or anesthesia. However, some handling and restraint procedures should be restricted to veterinary medical professionals due to the potential danger to the animal or handler. These require special skills, equipment, or facilities and possibly adjunct chemical restraint or complete immobilization by chemical restraint.

RESTRAINT OF INDIVIDUALS OR PORTIONS OF THEIR BODIES

- **Whole Body**
  - **Standing Restraint**
    - **Purpose:** Standing restraint of a cat is used for general physical examinations and occasionally intramuscular injections performed by another person.
    - **Method (*Procedural Steps 4.10*):**

Procedural Steps 4.10	Standing Restraint of Cats
1.	The handler puts the cat on a table and gains control of the cat's head by putting a thumb under the cat's collar or slip leash and cupping his fingers around the cat's chin and neck.
2.	At nearly the same time, he reaches over the cat with the other hand and lifts and hugs the cat's abdomen against his own abdomen so that the cat cannot sit.
3.	**Rectal temperatures** can be done by one handler with the cat in a standing position by holding the cat's body between the handler's left elbow and side with the cat's head directed toward the rear of the handler while the cat is restrained gently with the left hand holding the base of its tail. The lubricated thermometer is inserted with the right hand.

  - **Sitting Restraint**
    - **Purpose:** Sitting restraint position is often used for cephalic or jugular venipuncture and for oral administration of medications.
    - **Method:** This is done as with dogs, except that the head is restrained with a hand under the jaws and throat rather than an arm around the neck. The other hand gently pushes the rump down.
  - **Sternal Restraint**
    - **Purpose:** Sternal restraint is used for ear or eye treatments, administration of oral medications, and subcutaneous injections. It is also used for cephalic or jugular venipuncture.
    - **Method (*Procedural Steps 4.11*):**

Procedural Steps 4.11	Sternal Restraint of Cats
1.	The handler begins with the cat in standing position while restraining its front with the right hand under the neck and in front of the cat's chest.

Figure 4.6 Sternal restraint for cats.

Procedural Steps 4.11	Sternal Restraint of Cats
2.	The left hand is placed on the rump and pushes down gently while holding the front of the cat's chest.
3.	After the cat is sitting, the handler reaches around the left side of the cat with the left hand and grasps both front legs and gently slides them forward while pushing the cat's body down with his armpit and blocking lateral movement with his forearms (Figure 4.6).

- **For Subcutaneous Injection:** Restraint for subcutaneous injections can be sternal recumbency. The handling assistant presses down on the top of the cat's shoulders and lower neck while, at the same time, pressing the pelvis down with the other hand.
- **For Jugular Venipuncture:** If being held to have jugular venipuncture, the cat's front legs must be held over and below the level of the edge of the table. The edge of the table should be padded with a towel.
- **For Oral Administration** (*Procedural Steps 4.12*):

Procedural Steps 4.12	Sternal Restraint of Cats for Oral Administration
1.	Gentle cats can be administered oral medications if put in sitting or sternal position on a handler's lap with the cat facing the handler's knees.
2.	The handler's elbows and abdomen block sideways and backward movement.
3.	The handler's nondominant hand is used to restrain the cat's head and arrest forward movement, leaving the handler's dominant hand free to administer oral medications.
4.	If additional restraint is needed, the cat can be wrapped in a towel or put in a cat restraint bag.

- **Lateral Recumbency**
  - **Indication for Use:** Lateral recumbency should only be done when standing, sitting, or sternal restraint is not possible or desirable. It is the least comfortable and most intimidating for the cat. Its duration should be as short as possible.
  - **Method (*Procedural Steps 4.13*):**

Procedural Steps 4.13	Lateral Recumbency Restraint of Cats
1.	To restrain a docile cat in lateral recumbency, a handler begins with the cat in standing position.
2.	With his right arm, he reaches over the cat's neck and grasps both forelegs while placing a finger between the legs.
3.	With his left arm, he reaches over the cat's flank and grasps both hind legs with a finger between the legs.
4.	A finger between legs is very important for comfort of the cat and to maintain a grip on each leg.
5.	The cat is lifted while hugging it close to the handler's body so that it gently slides down on its side.
6.	The handler gently restrains the cat's head with his right forearm.

  - **Cat Restraint Bags:** Many cats will not tolerate being held in lateral recumbency. To prevent injury to the handler or the cat from struggling, lateral recumbency is often best done while the cat is in a cat restraint bag. Attempts to use lateral recumbency without a cat restraint bag will expose the handler's hands and arms to possible injury.
  - **Medial Saphenous Venipuncture and Other Uses:** Lateral recumbency is generally used to provide access to the medial saphenous vein on the inside of the hind legs. It is also used to administer intramuscular injections and perform toenail trimming in fractious cats.
- **Ventrodorsal Restraint**
  - **Indications:** A handler may hold the cat ventrodorsal (on its back) to do cystocentesis (urine collection by needle and syringe).
  - **Method:** If performed with the cat on a table, a soft, padded surface should be used such as a thick cushion or blanket. Alternatively, cats can be held ventrodorsal on a handler's lap.
- **Scruff-and-Stretch Hold**
  - **Indications:** If a cat is expected to strongly resist restraint, it can be restrained without a cat restraint bag for a short procedure by stretching it out.
  - **Method (*Procedural Steps 4.14*):**

Procedural Steps 4.14	Scruff-and-Stretch Restraint for Cats
1.	One hand grasps the cat's skin on the back of the neck, and although all fingers are used to grasp the skin, the thumb and index finger should gather the skin at a midpoint between the cat's ears.
2.	**Note:** Grasping the neck skin an inch or two in back of the ears leaves the head with enough mobility for the cat to turn its head and bite.
3.	The other hand grasps the back legs with the index finger between the legs.
4.	The cat is stretched out with its back slightly arched by pressing the shoulders and back with the heel of the hand holding the skin of the back of the neck (Figure 4.7).

Figure 4.7 Scruff-and-stretch restraint.

- **Criticism:** Scruffing is not universally accepted as an acceptable means of restraint, but it may be the best approach in some situations to protect the cat from greater injury or untreated disease, to reduce stress compared to alternative restraints, and to immediately protect handlers. In addition, scruffing does not impair the cat's ability to breathe, while wrapping a cat's neck or torso with a towel for restraint can cause respiratory difficulty. Scruffing should never be the primary means of restraining cats or used to lift a cat without another hand to support the body.
- **Proper Release:** Proper release of the cat from being scruffed is directly related to the ease or difficulty in handling the cat again. Release should only be done during a period that the cat is not struggling, followed by petting and offering small bits of food treats. If release is uneventful, subsequent handling may be easier rather than harder.

- **Restraint for Bathing**
  - **Wire Screens:** Domestic cats do not like being submerged or having running water on their body. If a wire screen is put in the bathing tub, cats with claws will often hold onto the screen rather than trying to jump out or climb up the handler's body.
  - **Mesh Bags:** Alternately, mesh restraint bags may be used to reduce the risk of scratches and the cat climbing on a handler during a bath. Regardless of the method used, cats protest with a mournful meow during a bath.
- **Head:** Restraint of a cat's head without risk of being clawed by either front or back claws requires a cat restraint bag, a cat declawed on all four feet, or a very tolerant cat.

  - **One-Hand Head Hold**
    - **Body Restraint First:** The one-hand head hold is used after the body is restrained. The most common reason for this hold is to administer oral medication.
    - **Method (*Procedural Steps 4.15*):**

Procedural Steps 4.15 One-Hand Head Hold of Cats	
1.	The handler's nondominant hand is used to hold the head.
2.	The head is grasped on top of the head with the middle finger and the thumb holding the zygomatic arches (cheek bones) and pressing the back molars area.
3.	This grasp permits restraint of the head while being able to press the jaws open with pressure on the back of the jaw with the nondominant hand and on the lower incisors with the dominant hand.

- **Two-Hand Head Hold** (*Procedural Steps 4.16*):

Procedural Steps 4.16 Two-Hand Head Hold of Cats	
1.	The handler grasps both sides of the cat's neck just behind the ears.
2.	Placing a towel over the cat's head first may allow the head to be grasped with less risk of being bitten.
3.	The handler's thumbs should be over the skull between the cat's ears, and his index fingers should be just behind the angle of the jaw and his fingers under the jaw bones, avoiding any pressure to the cat's throat.
4.	Control of side movement of the cat's body is attained with the handler's forearms.

RESTRAINT OF YOUNG, OLD, OR SICK/INJURED CATS

- **Kittens**: Queens should be separated and removed from the room if kittens will be restrained for examination or treatments. During the first 2 weeks of life, kittens cannot see, hear well, or control their body temperature well, so they must be carefully handled and kept warm.
- **Senior Cats:** Senior cats should not be handled for long periods, since they tire easily. If they are picked up, when they are released, they should be placed back on a floor gently.
- **Injured or Sick Cats:** Normally gentle, friendly cats will scratch and bite if they are in pain. Injured cats should be handled with towels to prevent a handler from being bitten. Crates or cardboard boxes are ideal for transport of injured or sick cats.

INJECTIONS AND VENIPUNCTURE

- **General Considerations:**
  - **Importance of Animal Restraint:** Insertion of transcutaneous needles for injection or aspiration in cats carries the risk of slashing tissue beneath the skin, including damage to nerves and blood vessels, and breaking hypodermic needles off in its body. The area in which the needle is to be inserted must be immobilized, and the cat's mouth and feet should be restrained from interfering with the procedure, especially venipunctures.
  - **Access to Veins:** Access to veins is needed for collection of blood samples and to administer medications intravenously.
    - **Use Minimal Restraint When Possible:** Restraint for venipuncture requires restraints that are relatively comfortable for the cat and do not make it feel trapped, e.g., it should not be squeezed when not resisting the restraint.
    - **Be Prepared to Control Resistance:** However, the method of restraint should allow the handler to immediately gain tight control if the cat resists. A sharp pain is often felt when the needle goes through the wall of the vein. Handlers must be mindful of this, because it is not the time for the cat to move and cause the venipuncture to fail.
  - **Coordinated Withdrawal of Needle:** It is best to coordinate withdrawal of the needle by the phlebotomist and release of the vein occlusion by the handler. Release of the vein occlusion should occur before the needle is withdrawn to

Figure 4.8 Cephalic venipuncture restraint.

reduce the leaking of blood from the venipuncture site. Compression should then be applied to the venipuncture site.

- **Cephalic Venipuncture**
  - **Body Restraint:** The cephalic vein is on the front of the foreleg. Sitting or sternal restraint is used to position cats for cephalic venipuncture (Figure 4.8).
  - **Method (*Procedural Steps 4.17*):**

Procedural Steps 4.17 Cephalic Venipuncture in Cats	
1.	If the cat is facing the handler's left side, the handler's left hand is placed under the cat's neck, and its neck and head are held close to the handler's body, while the right hand goes over the cat's body.
2.	The cat is held close to the handler's side with the handler's elbow.
3.	His right hand grasps the cat's outside elbow, with a middle finger or ring finger behind the elbow to keep the leg from escaping the hold.
4.	The thumb of the right hand holding the elbow reaches across the front of the elbow, squeezes the leg slightly, and then pulls the skin gently toward the outside, which occludes and stretches the cephalic vein to facilitate venipuncture by the phlebotomist.

- **Medial Saphenous and Femoral Venipuncture**
  - **Body Restraint:** The femoral vein is located superficially on the surface of a thigh, and one of its branches, the medial saphenous vein, is superficial on the medial surface below the stifle. Preferred restraint methods for femoral or medial saphenous venipuncture are to use the lateral restraint hold for docile cats and, for resistant cats, a towel wrap (burrito wrap) or a cat restraint bag.
  - **Method (*Procedural Steps 4.18*):**

Procedural Steps 4.18	Medial Saphenous or Femoral Venipuncture in Cats
1.	To perform a venipuncture on the medial saphenous without a cat restraint bag, a handler must restrain the cat in lateral recumbency.
2.	The phlebotomist grasps the paw of the lower hind leg and extends the leg.
3.	With one hand restraining the front of the cat, the handler's hand on the hind legs shifts to hold the upper hind leg in a flexed position, exposing the medial surface of the lower leg.
4.	The heel of the handler's hand on the upper hind leg can then press and occlude the femoral vein near the cat's groin for the phlebotomist to visualize the vein.

- **Jugular Venipuncture**
  - **Purpose:** Jugular veins are in the lower side of the neck on each side of the windpipe (trachea). The jugular vein is used to administer large volumes of fluids into the bloodstream or collect large volumes of blood. In adult cats, it should be reserved for critical care and emergencies. It is the preferred site for blood collection from kittens because of their size.
  - **Method (*Procedural Steps 4.19*):**

Procedural Steps 4.19	Jugular Venipuncture in Cats
1.	If a collar is present, remove it, because if it remains on and is pulled up out of the way, it will press on the upper aspect of the neck and prevent blood from entering the jugular veins where venipuncture is performed.
2.	The cat should be moved to the right edge of the table and placed in a sitting position or sternal recumbency.
3.	The handler holds the cat's front legs with his right hand (Figure 4.9).

Figure 4.9   Restraint for jugular venipuncture.

Procedural Steps 4.19	Jugular Venipuncture in Cats
4.	If the cat is in sternal recumbency, both front legs will have to be held over the edge of the table, which should be padded, and downward.
5.	It is very important to have control of the front legs, because some cats will reach up and push the phlebotomist's hand holding the syringe away, ruining the venipuncture
6.	The handler then holds the cat's head with his left hand on the lower aspect of the jaw and points the cat's nose toward the ceiling.
7.	A cat muzzle may aid in calming the cat and assist the handler's grip on the cat's head.
8.	The phlebotomist occludes the vein for the venipuncture.

- **Lateral Recumbency:** Lateral recumbency can also be used to access the jugular vein. Adequate restraint in cats for jugular venipuncture in lateral recumbency requires a towel wrap or a cat restraint bag.
- **Ventrodorsal Recumbency:**
  - Another restraint for jugular venipuncture is to place the cat in ventrodorsal recumbency (lying on its back) on a handler's lap with the cat's head toward the handler's knees. It also requires a towel wrap or a cat restraint bag.
  - **Method (*Procedural Steps 4.20*):**

Procedural Steps 4.20	Ventrodorsal Recumbency Restraint for Jugular Venipuncture in Cats
1.	The cat's hindquarters and pelvic area are restrained with the handler's left forearm and elbow while the right hand holds both forelegs with a finger in between.
2.	The left hand holds the neck, with the thumb used to occlude the jugular vein.
3.	The phlebotomist holds the cat's chin and head with the nondominant hand and performs the venipuncture with his dominant hand.

- **Injections**
  - **Intramuscular**
    - **Risks:** Intramuscular (IM) injections can cause serious injury. If the cat struggles during the injection, the needle slashes surrounding tissue. Therefore, a handler should not try to administer IM injections without assistance.
    - **Restraint:** The cat may be restrained in standing or lateral recumbency positions.
    - **Locations:** Intramuscular injections may be given into the caudal muscles of the thigh (semimembranosus and semitendinosus), cranial muscles of the thigh (quadriceps), or lumbar (lumbodorsal) muscles. Most IM injections are given into the caudal muscles of the thigh (Figure 4.10).

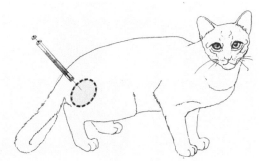

Figure 4.10 Intramuscular injection site in cats.

- **Immobilization:** Movement during the injection must be minimized to avoid needle trauma to the tissues, but the cat should not be squeezed with restraint if possible.
- **Subcutaneous**
  - **Method:** Subcutaneous (SC) injections can be administered under the skin of the upper half of the body in the caudal aspect of the neck to mid-thorax while a handler restrains the cat in sitting or standing restraint. Injections should not be administered on the dorsal midline. Fractious cats may require a scruff hold, scruff-and-stretch hold, or towel wrap, whichever is considered to be the safest with the least stress to the cat.
  - **Vaccine Sarcomas:** Because of the risk of vaccine-induced tumors in cats, the American Association of Feline Practitioners recommends that feline panleukopenia, herpesvirus-1, and calicivirus vaccines be given below the right elbow; rabies vaccine below the right stifle; and feline leukemia vaccine below the left stifle. These areas can be amputated if otherwise inoperable tumors occur. The risk of vaccine-induced tumors is less than one per 10,000 doses of vaccine (Figure 4.11).

Administration of Oral, Ophthalmic, Otic, and Transdermal Medications: Oral administration can be facilitated by training as a kitten. Giving small amounts of crunchy treats or pieces of kibble with a pill gun and liquid treats with a syringe for liquid medications can desensitize the cat to better tolerate their use to administer medications in the future.

- **Oral**
  - **General Considerations:**
    - **Difficult for Owners:** Cats are averse to oral administration and can be dangerous to handlers, especially owners, if not declawed.
    - **Adversity to Swallowing Tablets or Pills:** Cats will not swallow intact pills or capsules hidden in food treats.
  - **Tablet or Capsule Medication:**
    - **Hiding Medication:** Hiding solid medication for oral administration in food is always preferable to forcing oral medication into a cat's mouth (*Procedural Steps 4.21*).

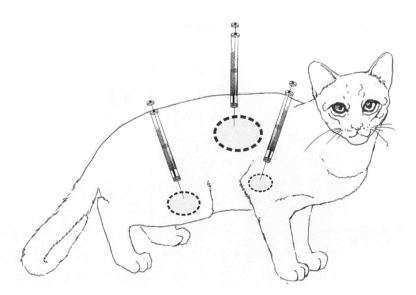

Figure 4.11   Subcutaneous injection sites in cats.

Procedural Steps 4.21	Hiding Tablets or Capsules in Food for Cats
1.	A small amount of a pasty, moldable cat-preferred treat should be offered, such as a paste of tuna, crushed crackers, and mayonnaise.
2.	If readily eaten, another small amount of a treat with hidden medication is offered.
3.	Capsules should be opened and sprinkled in the paste, and tablets should be crushed before adding to the paste until medication instructions prohibit it.
4.	After the cat consumes the medicated treat, it should be given a small portion of an unmedicated treat.
5.	If hiding the medication in food is not successful in medicating the cat, restraint for oral administration is necessary.

● **Direct Administration of Tablets or Capsules (*Procedural Steps 4.22*):**

Procedural Steps 4.22	Direct Administration of Tablets or Capsules to Cats
1.	Administering a capsule or tablet may require restraint using a restraint bag or towel.
2.	Sitting restraint is used for oral administration of medications.
3.	An assistant holds the cat's rump against his abdomen while gently holding the cat's forelegs to prevent forward movement and raising of the paws.
4.	Alternatively, a rolled towel can be wrapped around the front of the neck to block the front legs from reaching up and forward.
5.	The person administering the oral medication restrains the cat's head with the one-hand head hold.
6.	The cat's head is tilted back (Figure 4.12).

Figure 4.12   Restraint position for oral administration of solid medications.

7.	The tablet or capsule is held between the handler's right index finger and thumb, and the tip of the middle finger on the right hand is used to push the lower incisors down, while the left thumb and fingers put pressure on the cheeks to push them in, aiding opening of the jaw.
8.	The tablet or capsule is pushed into the mouth, and if necessary, the eraser end of a pencil is used to push the medication back over the hump of the tongue.
9.	A syringe of 2 ml of palatable liquid, such as tuna juice, should follow to reduce the chance of medication lodging in the esophagus and to countercondition the cat to oral administration restraint.x

- **Pill Gun:** Rather than using the eraser end of a pencil to push a tablet or capsule further down the throat, a pill gun can be used (*Procedural Steps 4.23*) (Figure 4.13).

Procedural Steps 4.23 Oral Administration of Tablets or Pills with a Pill Gun	
1.	These are similar to a syringe with a rubber end that entraps the tablet or capsule and a plunger to dislodge the medication after it is placed over the hump of the tongue, when the mouth is open (Figure 4.14).
2.	The gun should be rinsed and allowed to dry after each use.
3.	The cat's tolerance to pill guns may be improved by occasionally using the pill gun to deliver a special food treat.

- **Liquid Medication:** Any oral medication can be aspirated and cause respiratory problems, but the risk is particularly high with oral liquids. Oral liquid medications should never be administered to a cat with its mouth open. The medication should be aspirated into an oral syringe.
  - **Method** (*Procedural Steps 4.24*):

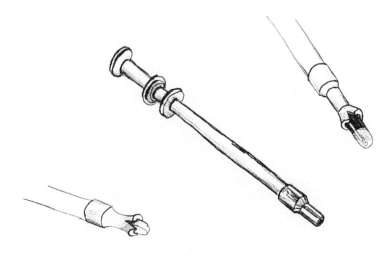

Figure 4.13   Pill guns can deliver tablets or capsules.

Figure 4.14   Oral administration using a pill gun.

Procedural Steps 4.24 Oral Administration of Liquid Medication to Cats	
1.	The cat's body should be restrained in the same manner as when administering solid medications.
2.	However, the mouth is held closed and the liquid medication slowly injected through the corner of the upper and lower lips, into the pouch between the cheek and teeth.
3.	By injecting slowly and holding the head pointed upward slightly, the medication will run around the dental arcade and be swallowed safely.

  – **Palatability:** Unpalatable liquids may be able to be diluted in beef or chicken broth, clam juice, or tuna juice to improve taste without altering the effects of the drug.
- **Ophthalmic:**
  - **Preferred Restraint:** To medicate a cat's eyes, a handler is generally needed to restrain the cat's head and body. The handler holds the cat in sitting position with its rump against the handler's chest or abdomen while holding the forelegs at the shoulder. The examiner holds the cat's upper and lower eyelids apart for an examination.
  - **No-Assistant Restraint:** Medication can be administered to a gentle cat by putting it in sitting position, facing away from the handler (***Procedural Steps 4.25***).

Procedural Steps 4.25 Application of Ophthalmic Medication to a Cat without an Assistant	
1.	The handler restrains the cat with his forearm and elbow pressing the cat against the handler's side.
2.	The hand of the restraining arm restrains the head.
3.	The middle finger of the treatment hand pulls the lower lid down, and the ointment or eye drop bottle is held with the index finger and thumb.
4.	The medication is dropped into the gap between the sclera and lower lid created by pulling the lower lid down.

- **Otic:**
    - **Body Restraint:** To medicate a cat's external ear, a handler is often needed to hold a cat in a sitting position and restrain its forelegs in order to administer ear medication. Gentle cats may be restrained in sternal position in a handler's lap for ear canal examination and administration of ear medications.
    - **Head Restraint:** The head should not be restrained by holding the ear flap. The cat's head should be restrained with a hand on the back of its neck and fingers and thumb behind the jaws. The head and neck are turned, as needed, for the ear canal exam or administration of medications. Fractious cats may need to be wrapped in a towel or placed in a cat restraint bag.
- **Transdermal:** The hairless interior surface of the cat's external ear can be used to topically apply some gel medications for transdermal administration. Required restraint is usually minimal.

CYSTOCENTESIS

Cystocentesis is collection of urine using a needle and syringe. Urine is obtained by puncturing the abdomen, entering the urinary bladder, and aspirating the urine. Restraint positions should be ventrodorsal or lateral recumbency. These positions permit stretching the cat's body while limiting torso movement.

SPECIAL EQUIPMENT

CAT RESTRAINT BAG

- **Canvas or Nylon:**
    - **Calming Effect:** Canvas or nylon cat restraint bags tend to calm cats after they are zipped in. Bags provide excellent restraint without the need to tightly restrict movement as with hand holds or towel wraps.
    - **Aids Venipuncture:** Zipped leg openings permit access to leg veins for venipuncture. Restraint by an assistant is minimal, and safety is high for cats and handlers. However, bags are for cats that may resent a procedure such as venipuncture, not overtly aggressive cats or cats already distressed.
    - **Method (*Procedural Steps 4.26*):**

Procedural Steps 4.26 Proper Use of a Cat Restraint Bag	
1.	Cats should be relatively quiet and handled gently to be placed in the bag properly.
2.	To put a cat in a restraint bag, the handler opens the bag and places it on a table.
3.	The cat is then placed on top of the bag.
4.	The neck portion of the bag is first clasped around the cat's neck, after which the bag is placed around the cat (Figure 4.15). It is important to zip the bag with one or two fingers under the zipper to prevent the cat's hair from getting caught in the zipper.

Procedural Steps 4.26	Proper Use of a Cat Restraint Bag
5.	Smaller zippered ports are present for access to legs (Figure 4.16).

- **Caution:** A cat in a bag should never be left unattended. Although once in the bag, a cat usually does not struggle to move, they should still be continuously supervised to ensure they cannot roll off a table.
- **Safe, Quiet Release:** After the procedure is completed, the cat should be slowly removed and petted in stages to desensitize them to the experience.

- **Mesh Bags:** Loose-fitting mesh restraint bags with a pull-string opening for minor restraint may be of value in handling cats during baths, brief transport, or administering medication.

Figure 4.15 Application of a cat restraint bag begins with fastening it around the neck.

Figure 4.16 Cat restrained in a bag with access to a hind leg.

MUZZLES

- **Types:** Leather, nylon, or cloth cat muzzles are open-ended muzzles designed to cover the eyes (Figure 4.17). Cloth muzzles pose less risk for injuring the eyes.
- **Purposes:** Cats are often quieted by being blindfolded with the muzzle. However, if the cat resists the muzzle, their feet must be well restrained. A one-hand hold restraint of the head may be easier to maintain with the traction provided by a muzzle.

GLOVES WITH GAUNTLETS

- **Purpose:** Leather gloves with gauntlets are excellent protection against scratches, but some cats can bite through them.
- **Bite Protection:** If there is danger of biting, the hand should be partially inserted in the glove. The empty fingers of the glove can then be offered as a distraction while the cat is captured by other means. However, leather gloves may carry stressful odors and exacerbate a cat's anxiety.
- **Desensitization to Pressure:** Gloves also desensitize the handler's hands to the pressure being exerted for restraint, which can contribute to the cat escaping restraint or being injured by excessive restraint.

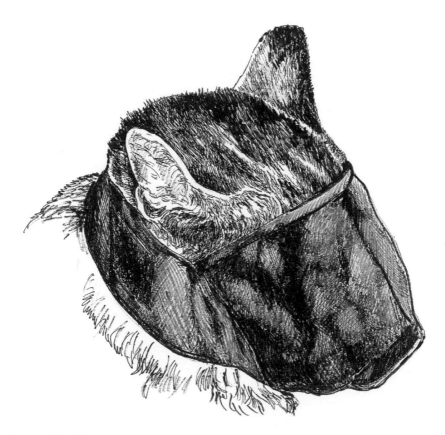

Figure 4.17   Cat muzzle.

LOOP ON A FLEXIBLE ROD

- **Description:** Leads with slip rings can be placed over the head to capture a cat by the neck to control the head while a hand is used to scoop up the body. A commercial loop on a flexible rod is made for capture of cats.
- **Use:** The loop on a flexible rod works well to quietly place a slip leash on a cat's neck and gently pull it snug on the neck, similar to the Mongolian uurga used to capture horses. The flexible rod should not be used to bend the leash and snap on the cat.

NETS

- **Opposing Nets on a Frame:** Opposing nets on metal frames attached with pivot points are available to restrain cats. They create a giant salad tong-like restraint tool that can be used to entrap a cat in a cage and remove it.
  - **Other Uses:** Opposing nets can also be used for transporting cats short distances if the cat's body is supported with one hand under the cat. In addition, the netting can restrain a cat for administering medications or chemical restraint.
  - **Effectiveness:** It is effective if the cat is entrapped in a standard cage at handler chest height. It is not effective for capturing cats in the open on the ground or a floor.
- **Hoop Nets:** Commercial hoop nets on long poles are available for the capture and restraint of cats.
- **Mesh Bags:** Nylon mesh laundry bags can also be used. The size of the mesh in the netting should be sufficiently small (one-fourth inches or less) to prevent cat entanglement and injury.

CAPTURE TONGS

- **Description:** Cat tongs are long-handled clamps for capture and restraint of vicious cats by the neck (Figure 4.18). The tongs are used to clamp the neck and restrain the head just before other means of restraint are applied.
- **Risks to the Cat:** The pressure on the neck, applied by tongs, is difficult to control when a cat is struggling, and if the body is not controlled at the same time the tongs are used, a struggling cat could break its neck.

CAGE SHIELD

- **Description:** A cage shield is a wooden or metal frame the same height and width of the inside of a cage, covered with mesh wire, and with a center brace and a centrally placed handle.
- **Use:** The shield is slid into a slightly open cage door and used to push a fractious cat against the back of its cage to administer chemical restraint.

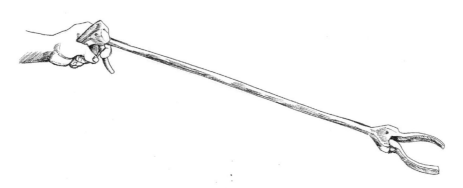

Figure 4.18   Cat capture tongs.

SQUEEZE CAGE

- **Description:** Wire cages are available with a sliding partition that permits a cat to be compressed inside a cage so that minor procedures can be performed or chemical restraint administered (Figure 4.19).
- **Use:** The use of squeeze cages should be restricted to the most vicious cats and preferably for one-time use on a cat.

HEAD MOVEMENT-LIMITING DEVICES

- **Elizabethan Collars:** E-collars are applied as with dogs (Figure 4.20). If tolerated by the cat, the collar will restrain a cat from chewing on its hindquarters. They can also provide the handler some protection from being bitten while handling or restraining the cat. Elizabethan collars must be removed to allow eating and drinking.
- **Globe Shield Collars:** Spherical semiopaque (globe-shield) collars that encapsulate the head and have an opening in front for breathing and vision are commercially available as a means to protect handlers from bites. Globe shield collars do not allow drinking or eating and must be frequently taken on and off, which could actually increase the risk of being bitten.
- **Broad-Band Cervical Collars:** Thick, broad collars wrap snugly around the neck like a human cervical collar to limit the cat's ability to reach areas of its body. A retention strap is run from the top of the broad band, around the lower aspect of the thorax and behind the front legs.

Figure 4.19   Squeeze cage.

Figure 4.20   Elizabethan collar.

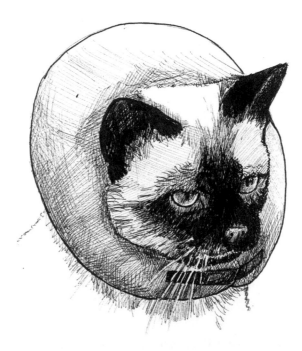

Figure 4.21   Pneumatic movement-limiting cervical collar.

- **Benefits:** Broad-band cervical collars permit better peripheral vision and hearing and do not catch on objects as does the Elizabethan collar. They also allow eating and drinking. However, access to most of the front legs is possible.
- **Construction:** Cervical collars for movement limitation can also be pneumatic (Figure 4.21). Heavy canvas construction is needed for protection from cat claws.

PINCH-INDUCED BEHAVIORAL INHIBITION

- **Description:** Two-inch paper clip binders, bent to reduce the pressure applied, have been recommended to be used on the loose skin of the upper neck as a *twitch* to distract difficult-to-handle cats in the same manner as scruffing.
- **Effect Theory:** It is theorized that the pinching simulates being carried as a small kitten by the queen. However, freezing from fear may be another possibility. Similar commercially produced clips for cat restraint are also marketed.

TRANSPORTING CATS BY AUTOMOBILE

- **Restraint:** If transporting a cat by car or in truck cabs, cats should ride in a back seat and be restrained in a crate that is strapped to floor anchors. This protects cats from airbag injuries and the driver from interference with driving (vision obstruction, interference with braking) or being injured by a cat becoming a missile in an accident. More information on travel restraints is available at the Center for Pet Safety: www.centerfor-petsafety.org/.
- **Preparation:**
  - **Trial Trips:** Cats should be desensitized to car travel by experiencing frequent short trips by car to a pleasurable destination with no adverse events during their primary socialization period (2 to 7 weeks of age), or as soon after that as possible.
  - **Adjustment to Carrier:** If cats will be transported in a carrier, pretrip training should include free access in and out of the carrier, being fed in the carrier, and sleeping in the carrier to develop a feeling of security while in the carrier.
- **Temperature Concerns:** If the temperature is over 72°F, cats should not be left in cars. At 72°F, the inside of a car can reach 100°F in 30 minutes. Temperatures below 55°F may be too low for some cats.
- **Interstate Travel:** Interstate travel must meet federal requirements. These include a certificate of veterinary inspection, provision of adequate shelter from all elements and protection from injury, sufficient cleanliness to avoid contact with urine and feces, protection against hazardous temperature extremes, uncontaminated and nutritious food at least once per day, and a program of parasite control.

NOTE

Additional recommended readings on cat handling are available in references on multiple species of small animals provided in the Appendix.

CAT HANDLING REFERENCES AND SUGGESTED READING

1. Babovic N, Cayci C, Carlsen BT. Cat bite infections of the hand: Assessment of morbidity and predictors of severe infection. J Hand Surg 2014;39:286–290.

2. Gruen ME, Thomson AE, Clary GP, et al. Conditioning laboratory cats to handling and transport. Lab Animal (NY) 2013;42:385–389.

3. Lantos PM, Maggi RG, Ferguson B, et al. Detection of *Bartonella* species in the blood of veterinarians and veterinary technicians: A newly recognized occupational hazard? Vector Borne Zoonotic Dis 2014:14:563–570.

4. Moody CM, Dewey CE, Niel L. Cross-sectional survey of cat handling practices in veterinary clinics throughout Canada and the United States. J Am Vet Med Assoc 2020;256:1020–1933.

5. Moody CM, Mason GJ, Dewey CE, et al. Getting a grip: Cats respond negatively to scruffing and clips. Vet Rec 2020;186:385.

6. Moody CM, Picketts VA, Mason GJ, et al. Can you handle it? Validating negative responses to restraint in cats. Appl Anim Behav Sci 2018;204:94–100.

7. Rodan L, Sundahl E, Carney H, et al. AAFP and ISFM feline-friendly handling guidelines. J Feline Med Surg 2011;13:364–375.

8. Scherk MA, Ford RB, Gaskell RM, et al. 2013 AAFP Feline Vaccination Advisory Panel report. J Fel Med Surg 2013;15;785–808.

9. Siven M, Savolainen S, Rantila S, et al. Difficulties in administration of oral medication formulations to pet cats: an e-survey of cat owners. Vet Rec 2017 Mar 11;180(10):250. doi:10.1136/vr.103991. Epub 2016 Dec 15.

10. Tuzio H, Edwards D, Elson T, et al. Feline zoonoses from the American Association of feline practitioners. J Fel Med Surg 2005;7:243–274.

5

OTHER SMALL MAMMALS

DOI: 10.1201/9781003110927-5

Table 5.1 Common Small Mammal Pets Other Than Dogs and Cats	
•	Mice
•	Rats
•	Hamsters
•	Gerbils
•	Guinea pigs
•	Chinchillas
•	Degus
•	Sugar gliders
•	African pygmy hedgehogs
•	Rabbits
•	Ferrets

In addition to dogs and cats, other small mammals are domesticated or tamed and kept as pets or used in research (Table 5.1). Ten percent of U.S. households have other small mammal pets, often referred to as *pocket pets*.

NATURAL BEHAVIOR OF SMALL MAMMALS

The natural behavior of small mammals depends on whether they are prey or predator. Most are prey and have similar characteristics. The only domesticated small mammal other than dogs and cats that is strictly a predator is the ferret. Rats are primarily prey, but in some situations, they can be predators.

GENDER INCOMPATIBILITIES

- **Problem:** Determining the gender of small mammals is important to prevent mixing groups that are incompatible. For example, if not raised together, adult male mice or male rabbits will fight each other, and adult female hamsters will fight each other.
- **Determination of Gender:**
    - **Difficulties:** Male rodents have open inguinal canals which enables them to pull their testes from the scrotum into the abdomen. Therefore, lack of external (scrotal) testes cannot be relied on as evidence the animal is a female.
    - **AG Distance:** Anogenital (AG) distance is the length from the urinary papilla to the anus. AG distance is longer in males than females, and the external genitalia in a male is somewhat circular, while a female's is more like a slit (Figure 5.1).
    - **Extrusion of the Penis:** Measuring the AG distance can be used to determine gender for most small mammals, except for guinea pigs and young rabbits. Sexing these animals requires gently pressing on each side of the genital orifice in an attempt to be able to extrude the penis, or not.

PREY SMALL MAMMALS

- **Hiding Defense:** All prey small mammals need a dark area to hide in. Hiding is an inherent need that helps relieve the stress of constantly being on guard for dominant aggression of other members of the group or for predators. This characteristic of seeking a hiding area makes it fairly easy for most of them to be coaxed into an opaque plastic cup or a small bin for capture and transport.
- **Special Senses**
    - **Vision:** The vision of prey small mammals is good for detecting movements but poor for detail.

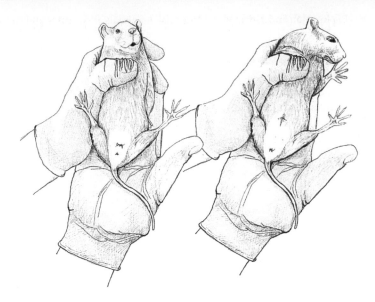

Figure 5.1   Determination of gender based on anogenital (AG) distance in rodents.

- **Hearing and Smell:** Prey small mammals have excellent senses of hearing and smell. Hearing is enhanced by also being able to detect sound vibrations from the ground. Handlers should wear plastic or rubber gloves to diminish the smell of predators that the handler may have recently touched or the smell of perfumes from hand soap.
- **Security of a Group:** With the exception of hamsters and hedgehogs, prey small mammals need the security of others of their species.
  - **Introduction of New Members:**
    - **Caution:** New members must be carefully introduced to an established group. Small mammals mark their territory and possessions by rubbing a part of their body on the object to be marked. Others with a different smell are rejected.
    - **Method (*Procedural Steps 5.1*):**

Procedural Steps 5.1	Introduction of Small Mammals to a New Group
1.	Introduction of a new member involves placing the new member in a nearby separate cage within sight and smell of the group.
2.	Over a period of days, the new member's cage is moved closer to the group cage.
3.	Next, the most docile member of the group is put into a third cage near the new member's.
4.	When the docile group member and the new member appear adjusted to each other, they exchange cages so that each acquires the smell of the other.
5.	After sufficient time to acquire the smells, the two should be able to be put in the same cage.
6.	If they accept the presence of each other, they can usually join the main group.

- **Physical and Mental Stimuli**
  - **Gnawing:** Since most small mammals instinctively have a need to gnaw, objects for gnawing (kiln-dried pine wood, cardboard tubes or boxes) should be offered.

- **Exploration and Movement:** Most small mammals also burrow and enjoy exploring PVC pipe in the size appropriately large enough for the species. Exercise wheels are popular enrichments for small mammal enclosures, but care must be taken that the construction of the wheel is safe for the small mammal species that will use it.
- **Light and Ambient Temperature:** Most small domestic or tame small mammals are nocturnal and unable to tolerate wide ranges of temperature. They should be allowed undisturbed rest during daylight hours in quiet surroundings. Their enclosures should not be exposed to drafts or direct sunlight.
- **Segregation from Predators:** The greatest stress a prey small mammal experiences is the sight, sound, or smell of a predator. All prey small mammals should be segregated, at all times, from dogs, cats, large birds, ferrets, and reptiles.

PREDATOR SMALL MAMMALS

Like prey small mammals, ferrets, which are predators, need social interaction with others of their species, long periods of undisturbed rest, and when awake, lots of opportunity for exercise and exploration. Much of the waking hours are spent in mock predator activities, such as wrestling with other ferrets, exploring burrow-like structures, and marking territory and possessions with their body scent.

SAFETY FIRST

HANDLER SAFETY

- **Bites:** All small mammals may bite when restrained. Rodents have incisors that angle backward. If bitten and the animal does not release its bite, the handler should replace it in its cage or box, where it will probably release its bite. Rabbits can bite and will do so on occasion.
- **Scratches:** Scratching from toenails on the hind feet is typically the greatest risk to handlers from rabbits.
- **Allergies:** Plastic gloves should be worn whenever handling rodents, because allergies to saliva, dander, or urine are common.

SMALL MAMMAL SAFETY

- **Tables for Handling:** All small mammals are handled on tables, and all will attempt to jump off, resulting in injury if restraint outside of the cage is not constant.
- **Tail Restraint Injury:** Mice, rats, and gerbils can be restrained by grasping the base of the tail, but they should not be held by the last two-thirds of the tail. Only the base of their tail can be safely held. Holding them by the distal aspects of the tail can allow rats to turn and bite the handler or gerbils to spin in an effort to escape and deglove (rip) the skin from the tail.
- **Minimize Duration of Restraint:** If capturing to perform a grooming or medical procedure, all materials should be readied prior to capture to reduce stress to the animal from prolonged restraint.
- **Avoid Stressful Odors:** Plastic gloves should be worn when handling nursing small mammals, especially rodents and rabbits. The young should be rubbed with used bedding after handling and before being returned to their enclosures. Otherwise, human scent may cause the mother to shun the babies.

KEY ZOONOSES

Apparently healthy captive-raised small mammals pose little risk of transmitting disease to healthy adult handlers who practice conventional personal hygiene. The risks of physical injury are greater than the risks of acquiring an infectious disease (Table 5.2).

Table 5.2 Diseases Transmitted from Healthy-Appearing Captive-Bred Small Mammals to Healthy Adult Humans

Disease	Agent	Means of Transmission	Signs and Symptoms in Humans	Frequency in Animals	Risk Group*
Bites	——	Direct injury	Bite wounds to face, arms, and legs	All small mammals are capable of inflicting bite wounds.	3
Rat Bite Fever	*Streptobacillus moniliformis* and *Spirillum minus*	Direct from bites or infected urine	Fever and joint pain	Uncommon in captive-bred rats	2
Pasteurellosis	*Pasteurella multocida*	Direct from rabbit or rat bites	Systemic illness signs (fever, inappetance), joint infections	Common in rabbits	2
Ringworm	*Trichophyton mentagrophytes*	Direct from contact with infected skin scale or hair and indirect from fomites	Scaly skin sores and hair loss with or without crusts or redness	Common in guinea pigs and rabbits	2
Lymphocytic choriomeningitis	*Arenaviridae* virus	Direct from body fluids or inhalation	Flu-like signs, stiff neck, abortion	Occasional presence in rats, mice, guinea pigs, and hamsters	4
Yersiniosis	*Yersinia enterocolitica*	Direct, fecal-oral	Bloody diarrhea	Occasional in guinea pigs	3
Cheyletiellosis	*Cheyletiella parasitovorax*	Direct and indirect with fomites	Itchy chigger-like bites on the skin	Common in rabbits	2
Salmonellosis	*Salmonella enterica*	Direct, fecal-oral	Diarrhea	Occasional in guinea pigs, sugar gliders, hedgehogs, and ferrets	3
Scabies	*Sarcoptes scabiei, Trixacarus caviae*	Direct from contact with infested skin	Itchy, chigger-like bites in the skin	*T. caviae* is common in guinea pigs, *S. scabiei*	2
Seoul virus	Hantavirus	Direct contact with body secretions, bites, or indirect from aerosolized dried feces	Fever, headache, back and abdominal pain	Carried by wild rats and rarely domesticated rats	2

*Risk Groups (National Institutes of Health and World Health Organization criteria. Centers for Disease Control and Prevention, Biosafety in Microbiological and Biomedical Laboratories, 5th edition, 2009.)
1: Agent not associated with disease in healthy adult humans.
2: Agent rarely causes serious disease, and preventions or therapy possible.
3: Agent can cause serious or lethal disease, and preventions or therapy possible.
4: Agent can cause serious or lethal disease, and preventions or therapy are not usually available.

SANITARY PRACTICES

- **Basic Practices:** Basic sanitary procedures are for small mammal handlers to wash their hands and to clean and disinfect tabletops and cages used in handling.
  - **Equipment:** Restraint equipment should be disposable or cleaned and disinfected.
  - **Attire:** A handler of small mammals should wear appropriate dress to protect against skin contamination with hair, skin scales, saliva, urine, and other body secretions. Use of plastic gloves in handling all small mammals is advisable.
  - **Personal Hygiene:** Basic sanitary practices should be followed, such as keeping hands away from eyes, nose, and mouth when handling small mammals and washing hands after handling the animals. Handlers should neither eat food nor smoke while handling small mammals.
- **Appropriate Housing and Handling:** The typical small mammal handler should only handle captive-bred, appropriately housed small mammals. Sick small mammals should be isolated from apparently normal animals. Rodent cages should be kept in a clean, well-ventilated area and never located in food preparation or eating areas.
- **Avoid Diseases of Small Wild Mammals:** Small mammals should be prevented from any direct or indirect contact with wildlife, particularly wild rodents. All food should be kept in rodent-proof containers. Wild rodent feces should be wiped with damp paper towels wetted in a solution of chlorine (1/4 cup bleach in a gallon of water).
- **Cleaning Containments:** While wearing gloves, handlers should clean small mammal enclosures and all enclosure contents on a regular basis (at least weekly). Gloves should be changed between cleaning separate enclosures. Enclosure and enclosure contents should be cleaned outside the primary family dwelling.
- **Preventative Medicine:** Ferret handlers should maintain current influenza vaccinations.

MICE

The mouse (*Mus musculus*), also called the house mouse, originated in Asia and has spread throughout the world. They were bred as pets in Japan during the 18th century.

- **Purpose:** Mice are common laboratory animals in biomedical research. They are also raised as household pets and food for captive reptiles and birds of prey.
- **Gender and Age Names:** Male mice are called *bucks*, female mice are *does*, and young mice are *pups*.

NATURAL BEHAVIOR

- **Aggressiveness:** Mice are more likely to bite than domestic rats and are not a good choice of pet for children. Adult males fight each other. Dominant males may chew the hair off some submissive members of their group, an activity called *barbering*.
- **Activity:** Mice are nocturnal and prefer to hide.
- **Special Senses:** The vision of mice is poor, but sound, smell, and movement are well detected. To attempt to avoid detection of their movement, cats remain still for long periods and advance slowly in spurts when stalking mice. Mice mark exploration paths with secretions from the soles of their feet and occasionally with urine and use their sense of smell to retrace paths.

APPROACHING AND CATCHING

- **Bite Caution:** Many mice will attempt to bite if unaccustomed to being restrained or restrained roughly. When they bite, they are often reluctant to let go.
- **Method (*Procedural Steps 5.2*):**

Procedural Steps 5.2	Approaching and Catching Mice
1.	Mice are best captured from a transport cage free of water and food bowls and exercise apparatus.
2.	All doors and windows in the handling room should be closed and possible hiding places blocked before removing a mouse from a cage.
3.	Mice may be grasped by the base of the tail with one hand and then scruff held by the other.
4.	Alternatively, they may be captured by providing a cup, called a rodent recreational vehicle (RV), for hiding and then entrapping them for transport.

- **Avoid Intimidating Odors:** Handlers should avoid carrying any sources of odor when handling mice, especially odors that might be associated with a predator.

HANDLING FOR ROUTINE CARE AND MANAGEMENT

- **Handling Tame Mice:** Frequently handled pet or research mice may be able to be handled by cupping hands around them (Figure 5.2).
  - **Method (*Procedural Steps 5.3*):**

Figure 5.2   Cupping a small rodent with gloved hands.

Procedural Steps 5.3	Basic Handling of Tame Mice
1.	Plastic gloves should be worn for protection from allergens.
2.	More often, a handler must grasp the base of the tail and lift the mouse to a surface that it can cling to, such as the handler's shirt or lab coat–covered arm or a small rug on a table.
3.	The mouse will continue to try to pull away while the handler continues to hold its tail.

- **Manual Transport:** Adult mice can be moved short distances by the base of tail, but if pregnant or obese, its body should be supported with the handler's other hand.
- **Full-Body Restraint Method:** Additional restraint can be applied by holding the tail and pressing the body down while grasping the skin on the back of the neck between the thumb and index finger (scruffing) and then swinging the body into the palm of the hand with the tail grasped between the ring and little finger (Figure 5.3). The nondominant hand should be used for restraint so the dominant hand can perform examinations, write notes, and administer medications.
- **Baby Mice:** Young mice, less than 2 weeks old, can be grasped by the loose skin of the neck and shoulder with a thumb and forefinger. Plastic gloves should be worn to prevent adding odor to the babies. Mother mice will cannibalize babies having strange odors.

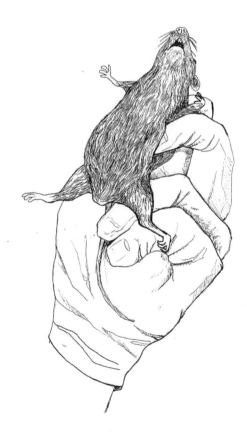

Figure 5.3   Scruff-and-stretch hold on a mouse.

Pet and research rats (*Rattus norvegicus*) were derived from the Norway rat (brown rat or wharf rat), which is the most common wild rat. They were bred as pets in Victorian England in the 19th century.

- **Stocks (Strains):** Sprague Dawley or Long Evans rats (specific families of *Rattus norvegicus*) are most easily handled and are preferred for pets and research.
- **Gender and Age Names:** Male rats are called *bucks*. Female rats are *does*, and young rats are *pups*.

NATURAL BEHAVIOR

- **General Behavior:** Rats are smart, interactive, nocturnal omnivores. They can be prey or predator.
- **Tame Rat Behavior:** If not overcrowded, they are clean and virtually odor free. They are less skittish than hamsters and gerbils, less likely to bite than mice, and less likely to scratch and injure a child than rabbit.
- **Genders:** Rats are similar to mice in that a large AG distance is consistent with being male. Males lack nipples, and their testicles are very pronounced at all ages. Rats tolerate living in mixed groups better than mice.
- **Aggressive Demeanor:** Aggressive rats will arch their back, fluff out the hair on their back, and swish their tail, similar to the aggressive posture of a dominance-aggressive cat.

APPROACHING AND CATCHING

- **General Approach:**
  - **Gentle Handling; No Scruffing:** Domestic rats are generally docile if handled gently and slowly. They resent being scruffed and are likely to bite if scruffing is attempted.
  - **Socialize Rat Pups:** Rat pups should be handled at weaning for socialization. Rats can be habituated to handling when young by *tickling*, a playful technique of desensitization to human hands.
- **Risk of Bites:** Rats should not be startled when they are asleep, or they may bite in defense. The primary defense response of rats is to hide when possible, but they will bite if cornered. The intent to bite is often signaled by a rat standing on its hindquarters and facing the approaching hand.
- **Preparations to Catch:** As with all small mammals, capture of rats is best attempted after removing feeders, water bowls, hiding boxes, or other moveable objects in the enclosure. All doors and windows should be closed and hiding places blocked before removing a rat from a cage.
- **Tail Restraint:** Capture should begin with grasping the base of the tail, but a rat should not be picked up by the end of its tail. Its body should be supported by the other hand (Figure 5.4).
- **Red Tears:** The Harderian gland is located behind the rat's eyeball. Stressed or sick rats will produce a red porphyrin from the Harderian gland that looks like blood, called red tears. Porphyrin can be identified by its fluorescence using a Woods light.

HANDLING FOR ROUTINE CARE AND MANAGEMENT

- **Minimal Restraint Techniques:** Rats like to hide, so many rats will become calm if allowed to hide in a coat pocket. They are more comfortable if allowed to move around on a sleeved arm and intermittently repositioned by grasping the base of the tail or the shoulders. Older rats often have chronic respiratory disease and can be severely stressed by restraint.
- **Total-Body Restraint Method (*Procedural Steps 5.4*):**

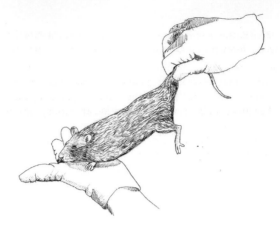

Figure 5.4   Restraint of a rat by the base of the tail and support of the body.

Figure 5.5   One-handed rat restraint.

Procedural Steps 5.4 Total-Body Restraint of Rats	
1.	Firm manual restraint of rats is performed by grasping the base of the tail with one hand and the rat's shoulders with the other hand, using the thumb under a foreleg and the jaw.
2.	The handler's index and middle fingers restrain the foreleg on the other side of the rat (Figure 5.5).
3.	The rat's chest or trachea should not be squeezed, because if its breathing is impaired, it may panic and attempt to bite when held or when released.

The common pet hamster (note: no "p" in hamster) is the golden (Syrian or teddy bear) hamster (*Mesocricetus auratus*), which originated in the desert region of Syria.

- **Physical Characteristics:** Hamsters have extremely loose skin, virtually no tail, hairless feet, and large cheek pouches for storing food or hiding and transporting valued possessions, including baby hamsters. Golden hamsters have paired glands in their skin over their flanks. The flank glands, also called hip spots, are more prominent in males, which use the gland secretions to mark their territory.
- **Types:** Unrelated hamsters to the golden hamster are the Russian hamsters (*Phodopus spp.*) and Chinese dwarf hamsters (*Cricetalus griseus*), which are smaller and less common than the golden hamster. Russian hamsters are usually brown or gray and more social and less nocturnal.
- **In Research:** Hamsters have been used for research on ear diseases, and because of their very short gestation period, they are used for teratogenic studies. They are less affordable for research, since they need to be housed individually. Consequently, much more space is required for cages than with other rodents.
- **Gender and Age Names:** A male hamster is called a *buck*. A female is a *doe*, and young hamsters are *pups*.

NATURAL BEHAVIOR

- **Common Habits:**
  - **Solitary Existence:** Adult hamsters prefer to live a solitary existence except at breeding. Hamsters are burrowing, nocturnal, desert rodents that are drowsy in the daytime and have poor eyesight, particularly in bright light. They arc likely to bite, especially if startled, and are not recommended as pets for children.
  - **Pseudohibernation:** In their natural habitat of the desert, adult hamsters primarily live alone in tunnels for cooler temperature and higher humidity than that on the desert's surface. They tolerate cold temperatures well and will go into pseudohibernation if the temperature goes below 48°F and could be mistaken as dead during this time.
  - **Danger of Falls:** In their natural habitat, hamsters have not had to evolve with an ability to negotiate cliffs and ledges well. In captivity, they are more likely to fall off exam tables than other small mammals.
  - **Food Hoarding:** They forage for food at night and then carry it back to their burrow in their cheek pouches. Pet hamsters also like to hide food. Bedding and other areas within the enclosure should be routinely checked for stashed rotting food.
- **Determination of Gender:**
  - **Method:** Sexing of hamsters is similar to the rat. The adult male's body protrudes more caudally due to the testicles, and the difference is striking. The AG distance is much longer in males. In dwarf species of hamsters, males possess a prevalent scent gland on the midline of their abdomen.
  - **Need to Know:** Male hamsters are much more docile than females. They should be separated after they are 6 to 10 weeks old. Females are larger and dominant to males, more aggressive, and more likely to fight, especially if pregnant or lactating.

APPROACHING AND CATCHING

- **Approach**
  - **Arouse if Sleeping:** If attempting to sleep, hamsters bury their head under their abdomen, which impairs their ability to see, smell, or hear a handler

approaching. A handler should be sure the hamster is awake before attempting to capture it to prevent startling it. To awaken a hamster, the handler should talk to it or jiggle its cage.

- **Separate Females from Litters:** Adult female hamsters are usually crankier than males. A nursing female should be captured when away from her litter. Female hamsters can be extremely aggressive when they are nursing.

- **Capture**
  - **Prepare Enclosure:** Capture is best attempted after removing feeders, water bowls, hiding boxes, or other moveable objects in the enclosure.
  - **Routine Capture:** Gentle hamsters can be captured by cupping with both hands and then supporting their body in the handler's palm or grasping the skin on the back of the neck. They also may be able to be induced to enter a small can or cup and moved after covering the opening.

HANDLING FOR ROUTINE CARE AND MANAGEMENT

Whole-Body Restraint: For the best physical restraint, the handler must do a full-body scruff hold (Figure 5.6).

- **Method:** The hamster is covered with one hand while pinning the head between the thumb and index finger, then without releasing the skin behind the neck, it is grasped with thumb and index finger and the skin of the back with other fingers and heel of the hand.
- **Caution:**
  - **Risk of Eye Injury:** Caution is required to not gather the skin too tightly near the head. Too much tension on the skin around the eyes can cause a prolapse of an eyeball (proptosis).

Figure 5.6   Full-body scruff hold on a hamster.

- **Risk of Aspiration:** When possible, restraint should be avoided if the cheek pouches contain food. Hamsters may aspirate cheek pouch materials if scruffed with their pouches full.

GERBILS

Gerbils (*Meriones unguichlatus*), also called *jirds*, originated from the deserts of eastern Mongolia and northeastern China.

- **Characteristics:** They have long-haired, thin tails. Most are sand colored with white underbellies. Both genders have a ventral marking gland on their abdomens that is used to mark their territory.
- **Use in Research:** Gerbils have been used in medical research on strokes in humans.
- **Biopollution Risk:** Gerbils are illegal as pets in California due to environmental risks from escape into the wild.
- **Determination of Gender:** In gerbils, the AG distance in males is a half inch, and the scrotum should be apparent after about 6 weeks. Only females have nipples.
- **Gender and Age Names:** A male gerbil is called a *buck*. A female is a *doe*, and young gerbils are *pups*. A group of gerbils is a *horde*.

NATURAL BEHAVIOR

- **Gentle, Gregarious, and Monogamous:** Gerbils are monogamous, and pairs should not be housed separately. They live best in small groups or pairs. Gerbils are not strongly nocturnal and are usually active during the day if other activities are going on. They are gentle and rarely bite, except with new cage members if introduced too rapidly to an established group.
- **Common Habits:**
  - **Quiet:** Gerbils are quiet and not as vocal as other small rodents (mice, rats, hamsters).
  - **Burrowing:** They like to burrow in sandy soils and should be provided deep bedding with solid-bottomed cages.
  - **Drink Little Water:** Having evolved in deserts, gerbils consume little water and produce small amounts of urine. They therefore do not produce strong urine odor characteristic of other rodents.
  - **Hide and Play:** Gerbils enjoy hiding and running through tunnels and using exercise wheels, which should have fine mesh to protect their toes.
  - **Gnaw:** They like to chew and should not have access to plastic objects due to risk of gastrointestinal obstruction from plastic pieces.
  - **Dust Baths:** Like other desert animals, they enjoy dust baths.
  - **Warning Thumps:** If stressed, gerbils will signal danger to others by thumping a hind foot.
  - **Aggression toward Weanlings:** After about 5 weeks from giving birth, it is recommended to remove the pups from the pregnant mother, because she may become aggressive toward the older pups.
- **Marking Gland:** Male and female gerbils have an oval-shaped ventral marking gland on their abdomen. The gland is more pronounced in postpubertal males, which use the gland to mark their territory.
- **Tear Stains:** Overcrowding, high humidity (more than 50% for gerbils), or other stresses in gerbils will cause porphyrin secretion in their tears. Porphyrins in tears will

stain skin and hair around the nostrils and eyes a reddish-brown and irritate the skin, progressing to skin sores.

APPROACHING AND CATCHING

- **Routine Approach and Capture:** Gerbils are seminocturnal desert animals that are easy to handle (Figure 5.7). If given time to adjust to the handler's smell and voice, they may climb into his cupped hands, if a handler's movements are appropriately slow.
- **Capture of Untamed Gerbils:** Gerbils without good handling experience may be captured with both hands or grasped in one hand using the methods used for mice, i.e., carefully grasping the base of tail or scruffing the neck and shoulder skin.
- **Avoid Distal Tail Restraint:** Do NOT capture gerbils by their tail, except carefully at its base. The skin will easily be stripped off during its struggling, an injury called *tail slip*.

HANDLING FOR ROUTINE CARE AND MANAGEMENT

- **Firm Restraint:** If firm manual restraint is necessary, a handler should grasp the body over the back with its head between the thumb and index finger or alternatively grasp (scruff) the skin on the neck and back and trap the tail with his little finger as done when restraining mice.
- **Risk of Seizures:** Handling duration should be kept as brief as possible. Gerbils become stressed from handling, and many will seizure if the stress is prolonged.

Figure 5.7 Gerbils do not usually resist gentle handling.

Guinea pigs (*Cavia porcellus*), also called cavies, are crepuscular, docile, and social animals which were domesticated as a food source about 7,000 years ago. They are from the Andes Mountains in South America, not Guinea.

- **Characteristics:** Guinea pigs squeal like pigs and have the same general body shape. They are related to chinchillas and have no tail and no hair on their ears.
- **Determination of Gender:**
    - **AG Distance:** The AG distance in guinea pigs is a marker for gender. However, the distances are not as distinguishable as with other rodents.
    - **Female Sexual Phenotype:** A female has a Y-shaped opening made by the close proximity of the vulva to the anus.
    - **Male Sexual Phenotype:** The male is characterized by gently pressing the abdomen in order to cause the penis to emerge, male nipples are less developed, and the testicles are evident in a mature male.
- **Gender and Age Names:** A male guinea pig is called a *boar*. A female is a *sow*, and young guinea pigs are *pups*.

NATURAL BEHAVIOR

- **Short Legged, Social, and Dominant Males:** Guinea pigs are rotund social rodents with short legs, small, hairless ears, and no tail. Adults weigh about 30 to 35 ounces. In the wild, they live in colonies (clans) of 5 to 10, with a dominant male, in burrows or crevices in rocks.
- **Common Habits:**
    - **Vocal:** Guinea pigs are vocal.
        - **Whistles, Chirps, Drilling Sounds:** They may whistle if alarmed or if greeting a known handler who feeds them, chirp if content, and make guttural drilling sounds if agitated by a perceived threat or from pain.
        - **Purrs:** They purr when content and feeling secure.
        - **Chattering and Hissing:** Teeth chattering and hissing are signs of irritation and possible aggression.
    - **Male Aggression:** Sexually intact males will challenge each other until dominance is achieved by one. True fighting and injury of an opponent is not common. Head-butting is a show of dominance, invitation to play, or irritation with the current situation.
    - **Freeze and Stampede:** They tend to freeze when startled and then scatter frantically. Panicked guinea pigs will stampede and injure smaller, weaker members of a group. A frantic attempt to escape can also lead to injury from falling if in an elevated cage or on a table.
- **Marking Glands:** Guinea pigs have special sebaceous glands in their skin on the top of their body and in the anal area which are used to mark their territory and possessions.
- **Special Senses:** They have good peripheral vision, typical of prey animals, and very good hearing with a frequency range up to 30,000 Hz (20,000 Hz is the upper limit for humans).

APPROACHING AND CATCHING

- **Defense Tactics of Guinea Pigs:**
    - **Squeal, Run, Hide, Freeze:** Guinea pigs are easily alarmed and will squeal loudly and attempt to evade capture by a stranger. Their primary means of defense is to either run or freeze in place.
    - **Low Risk of Biting and Scratching:** Some will bite if restrained, but their mouth is too small to inflict severe bites to an adult handler. They have four claws on each of the front feet and three on both rear feet.

- **Low-Stress Approach:**
  - **Best Approach:** An initial attempt to capture a guinea pig is best done using food as a lure or stroking its head and nose until calm, then grasping it with one hand underneath its chest and cupping the other under its rump. Young, small guinea pigs can be grasped with one hand.
  - **Do Not Scruff Hold:** Guinea pigs have little loose skin over their neck and shoulders, and attempts to scruff them can be painful to them.
- **More Assertive Approach:**
  - **Front and Back:** Another capture method is to use one hand from above to cover its head, blocking its vision, while covering the rump with the other hand and then reaching under it from behind. To pick up the guinea pig, the front hand is placed under the guinea pig's chest and the other hand under its rump.
  - **Provide a Hiding Opportunity:** When being carried, they will relax if allowed to hide with their body supported by the handler's forearm and their head in the crook of his elbow.

HANDLING FOR ROUTINE CARE AND MANAGEMENT

- **Routine Restraint:**
  - **Two-Handed Restraint Necessary:** Restraint of adult guinea pigs should always be two-handed (Figure 5.8). The thorax should be grasped either dorsally or ventrally with one hand as the other hand supports the guinea pig's rump.
  - **Never Restrain by Scruffing:** They should never be scruffed. Their body weight in comparison to their musculoskeletal system strength is too great, and back or neck injuries can result.

Figure 5.8 Restraint of guinea pigs must be two-handed.

- **More Assertive Restraint:** If there is resistance to restraint, the handler should grasp it without hesitation around the shoulders with one hand. Lift it primarily with the thumb under a leg and under the jaw (to block the animal from lowering its head to bite) and first two fingers around the shoulders without squeezing the thorax. Place the other hand under the body.
- **Restraint for Trimming Toenails:** A guinea pig's toenails need to be trimmed every 8 to 12 weeks. To assist in toenail trimming, a handler supports the guinea pig against his chest and holds behind the front legs with one hand and the other hand cupped under the rump. This has the guinea pig positioned in a C position.

CHINCHILLAS

Chinchillas (*Chinchilla lanigera*) are soft, odorless rodents and rarely bite. They are indigenous to the Andes Mountains of South America at 10,000 to 15,000 ft, living in groups of up to 100 individuals.

- **Characteristics:** Chinchillas have a rounded body, large, mouse-like ears, a long, furry, squirrel-like tail, and short legs. Chinchillas have large ears with a similar structure and range of hearing to humans. Because of this, they have been used as an animal model for human ear research.
- **Determination of Gender:** Chinchillas are sexed using the AG distance. A female has a urethral cone that resembles a penis, but it sits directly in front of the anus. In the male, there is bare skin between the anus and the urethral opening.
- **Gender and Age Names:** Chinchillas are simply referred to as a male and female. Young chinchillas are **kits**.

NATURAL BEHAVIOR
- **Effects of Extremely Dense Haircoat:**
  - **Impenetrable to Water:** Chinchillas are nocturnal relatives of guinea pigs and have extremely dense, soft fur (Figure 5.9). Their fur has about 50 to 60 hair shafts per follicle, compared to 10 to 15 in most dogs and 1 per follicle in humans. Chinchillas clean themselves by fine dust bathing.

Figure 5.9 Chinchillas have very dense hair that can be easily pulled out during some restraints.

- **Fur Slip Defense:** *Fur slip* , tufts of hair that are pulled from their follicles, is their primary means of defense after hiding. Other means of defense are bluffing by standing on their hind legs, chattering, barking, spitting, and urinating directly at their perceived threat.
- **Tolerant of Cold Weather:** Their dense coat provides good protection against cold even below freezing, but temperatures above 80°F (27°C) can cause heat stroke. Their broad ear flaps aid in dissipating heat.
- **Good Jumpers:** Chinchillas have strong back legs and are very good at leaping. Young ones can jump more than 6 feet high and have no fear of heights, but this characteristic puts them at high risk for falling injuries.
- **Squirrel-Like Eating Posture:** Chinchillas sit on their rump and eat using their forepaws.
- **Cecotrope Eaters:** They eat primarily at night and pass most of their feces at night. Like rabbits and some other rodents, they eat cecotropes (special nutrient feces) in the mornings to supplement their nutrition.

APPROACHING AND CATCHING

- **Approach:**
  - **Risk of Fractures:** They are curious animals that, with patience, can often lead them to be easily captured, although they are quick and can jump distances several times their body length. Care must be taken not to startle them, since they may leap hazardously and fracture their back.
  - **Raisins as a Lure:** Chinchillas love raisins, which can be used as a lure to catch them. However, more than two raisins in a day may cause diarrhea.
  - **Minimize Handling:** They usually like to be petted, but even well-tamed chinchillas do not relax when being held. If feeling threatened, they will try to urinate on a handler.
- **Capture**
  - **Avoid Fur Slip:** Grabbing their haircoat will cause fur slip. Fur slip requires at least 6 to 8 weeks to regrow and may come back a different color.
  - **Cupping Socialized Chinchillas:** Previously handled chinchillas can be cupped with both hands and swept toward the handler's body.
  - **Whole-Body Restraint:** For greater restraint, the handler should grasp the shoulders with one hand and hold the hind legs with the other hand in the same manner as holding a guinea pig.

HANDLING FOR ROUTINE CARE AND MANAGEMENT

- **Usual Manual Restraint:** The handler can hold the base of the tail to prevent jumping and support the body with the other hand. Holding the base of the tail to prevent jumping while supporting the body is acceptable. They should never be lifted by the tail or ears or scruffed.
- **Exam Tables:** Examination table surfaces should be covered by a towel or other nonslip material.
- **Towel Restraint:** Further restraint can be achieved by wrapping in a small towel.

DEGUS

The degus (DAY-goo; *Octodon degus*) is from semiarid regions of north central Chile. Their ears are large to aid in heat dissipation. They are also called *Chilean squirrels* and *brush-tailed rats*.

- **Characteristics:** They hold food with their forepaws, similar to North American squirrels. Degus are related to guinea pigs and chinchillas, have long-haired tails, and look like large gerbils.

- **Use in Research:** They have been used as a research model for development and aging.
- **Biopollution Risk:** Degus cannot be kept legally as pets in California, Georgia, and Alaska. Pet shops that carry degus for sale and breeders of degus must be licensed by the USDA.
- **Determination of Gender:** Female degus do not have a gap between the anus and urinary cone, while in males, there is a gap. The urinary cone is not the penis.
- **Gender and Age Names:** There is no special name for degus genders. A male degus is a male, and a female is a female. Young degus are *pups*.

NATURAL BEHAVIOR

- **Social, Diurnal, and Coprophagic Jumpers:** Degus are social, diurnal (active during daytime) rodents that are coprophagic. They are good jumpers.
- **Burrow and Enjoy Dust Baths:** They like to live in groups in burrows and clean themselves with dust baths, as do chinchillas.

APPROACHING AND CATCHING

- **Minimize Handling:** Although active, curious, and willing to approach humans, degus do not like a lot of handling.
- **Two-Hand Restraint Required:** They should be picked up with one hand supporting the hindquarters and the other supporting the thorax just behind the forelegs. They should never be picked up by their tail due to risk of degloving injury.

HANDLING FOR ROUTINE CARE AND MANAGEMENT

If socialized as juveniles, degus can be handled in the same manner as guinea pigs (Figure 5.10). Restraint by wrapping with a towel may be used if necessary.

Figure 5.10 Degus are handled similarly to guinea pigs.

- **Characteristics:** Sugar gliders (*Petaurus breviceps*) are small (5 to 7 inches long) tree-living (arboreal), nocturnal marsupials from Tasmania, Indonesia, New Guinea, and eastern Australia. Females have pouches on their abdomens. They have 40 to 46 teeth, including two large incisors that point forward, which are used to penetrate the bark of trees.
- **Biopollution Risk:** They are illegal in California, Pennsylvania, Hawaii, Alaska, and New York City to prevent biopollution.
- **Gender and Age Names:** There are no special gender names for sugar gliders. They are called male and female sugar gliders. Young sugar gliders are *joeys*.

NATURAL BEHAVIOR

- **Territorial:**
    - **Rejection of New Members:** They are very protective of their territory, and new sugar gliders not marked by the colony's dominant male will be attacked.
    - **Closed Colonies:** In the wild, sugar gliders live in colonies of 10 to 15 individuals in trees, nest in hollows of trees, and feed on insects and plant nectars.
    - **Marking Odors:** Males have a bald spot, which is a scent gland, on their heads. Males use their head glands and similar scent glands on their sternum and near their cloaca to mark other gliders and their territory. If frightened, male and female gliders will express their paracloacal scent glands. The secretion has a spoiled-fruit odor. They also mark their territory with their urine.
- **Gliding:** They are able to glide more than 100 feet between tree limbs by spreading their legs and using a fold of skin, the *patagium*, between their front and rear legs like a parachute and using their tail as a rudder (Figure 5.11).

Figure 5.11 Sugar gliders can glide long distances by spreading their patagium.

- **Dexterous:** They have five digits on each foot. The inner digit on the hind feet is bulbous, without a claw and opposing to the others like a human thumb, and allows them to easily grasp limbs. The second and third digits on the hind feet are fused together and aid the glider in grooming its haircoat.
- **Fearful of Raptors:** Their principal predator in the wild is the owl. As a result, sugar gliders become very stressed when in the presence or sound of birds.
- **Vocal:** Vocalizations include chattering for attention, yapping like a small dog when alarmed, and high-pitched crabbing when startled.

APPROACHING AND CATCHING

- **Socialized Gliders:** Sugar gliders should not be awakened to be caught, as this causes them significant stress. When awake, those that have been socialized to humans as juveniles may stay in a handler's open hand (Figure 5.12). If they must be caught, they can vocalize many different sounds when disturbed, and they can inflict a bite or gouge with their incisors.
- **Assertive Approaches:**
  - **Manual and Cloth Bags:** The neck can be grasped with the handler's thumb and middle finger and the index finger on top of the head. Cloth bags can be everted over the handler's hand. After grasping a glider, the cloth is folded back over the glider, leaving its head exposed.
  - **Towels:** Towels can be useful aids in capturing gliders, but looped cloth should not be used, nor cloths or towels with loose threads. Sugar gliders' feet can easily be caught in loops or loose threads, causing serious injuries.
  - **Sudden Increase in Lighting:** Increasing the light in the capture room or using a flashlight may cause them to freeze long enough to use a towel for capture.

Figure 5.12   Sugar gliders can be handled if socialized early and handled quietly.

HANDLING FOR ROUTINE CARE AND MANAGEMENT

- **Scruffing Restraint:** Safe physical restraint of sugar gliders is difficult. Scruffing the loose skin of the back can be done for restraint, although this will elicit loud crabbing from the glider.
- **Chemical Restraint:** Most procedures require chemical restraint.
- **Toenails and Looped Fabric:** Care should be taken to prevent their claws from getting caught in the handler's clothing fabric. If a toe is caught, freeing their toe may inadvertently injure their toe, wrist, or ankle. Their toenails may need to be trimmed periodically.

AFRICAN PYGMY HEDGEHOGS

African pygmy hedgehogs (*Atelerix albiventris*), also called the four-toed or white-bellied hedgehog, are solitary, territorial, nocturnal, insect-eating mammals from the southern Sahara desert that prefer to live alone.

- **Characteristic Spines:** Most of their body is covered by ¼- to 1-inch long spines (quills). The spines are used for defense and to cushion falls. African pygmy hedgehogs are not able to fling their spines.
- **Biopollution Risk:** Foot- (or hoof-) and-mouth disease is a viral disease of cloven-hoofed farm animals that has been reported in hedgehogs. Imported hedgehogs can also carry anthrax and could survive as feral animals after escape or abandonment in the southern U.S. As a result, African hedgehogs cannot be imported legally, and they cannot be legally owned in some states (California, Hawaii, Arizona, Georgia, Pennsylvania, Maine, and Vermont) and several cities.
- **Determination of Gender:** Females have five pairs of nipples and a very close AG distance. The male's AG distance is much larger, and the penis is located near the mid-abdomen.

NATURAL BEHAVIOR

- **Solitary, Grunting Foragers:** Hedgehogs in the wild live in a variety of environments, including in rock crevices, brush, or burrows. They are solitary except at breeding time.
- **Special Senses:** They are sensitive to strange sounds and have an excellent sense of smell, which they use in foraging for food, but their vision is weak. Tactile sensations are perceived by touching with their spines and vibrissae (whiskers).
- **Vocalizations:**
  - **Grunting:** Hedgehogs grunt when foraging for food. This hog-like vocalization and a preference to forage along hedgerows were how the small marsupial got its common name.
  - **Other Vocalizations:** Typical vocalizations are grunting, clicking, snorting, and sniffing, but hissing will occur if a hedgehog feels threatened. Screams occur if distressed.
- **Defense Tactics:**
  - **Curling into a Ball Position:** When feeling endangered, they will elevate spines on their forehead and curl into a ball.
  - **Escape:** They are good at digging, climbing, and swimming.
  - **Venom Resistance:** They are resistant to many venoms, including those of many snakes, bees, beetles, and spiders.
  - **Bites:** They have small teeth and will bite if irritated or threatened.

APPROACHING AND CATCHING

- **Adaptation to Handling:**
  - **Prime Time to Socialize:** Socialization of hedgehogs with handlers is best begun when the hedgehog is 6 to 8 weeks of age.
  - **Avoid Odors on Hands:** Handling young hedgehogs with bare hands will accustom them to the handler's odors. Use of perfumed hand soap or lotions should be avoided.
  - **Treat Rewards:** Additional positive conditioning to be handled can be provided with treats, such as mealworms, while the hedgehog is being handled.
- **Protect Skin of Hands:** Latex or light leather gloves or a towel should be used to handle strange or untrained hedgehogs. Although their spines are not barbed, spines may penetrate a handler's skin. When excited, they *anoint* (spread a thick, frothy saliva) their spines, which can cause skin irritation in some handlers.
- **Scooping Capture:** Capture of socialized hedgehogs requires slowly scooping them from underneath its belly with one or two hands (Figure 5.13). Their bellies are covered with soft fur, but their backs are covered with short, prickly spines. Gentle handling is needed to prevent them from rolling into a defensive ball and making a hissing sound.

HANDLING FOR ROUTINE CARE AND MANAGEMENT

- **Techniques:** Properly socialized hedgehogs can be held in cupped hands. Difficult hedgehogs can be scruffed by the skin between the ears. Alternatively, a rear leg can be grasped.
- **Avoid Uncurling Attempts:** Attempts to forcefully uncurl a hedgehog that has rolled into a defensive ball should be avoided due to the risk of injury to the hedgehog.

Figure 5.13   African pygmy hedgehogs require minimal restraint.

Rabbits (*Oryctolagus cuniculus*) captured on the Iberian Peninsula of Europe were domesticated by the Phoenicians about 3,000 years ago. Today's domesticated rabbits are descendants of the European rabbit, not the American cottontail rabbit.

- **Popularity:** Rabbits are the most popular other small mammal pets in the U.S. They are also the most common pet relinquished to animal shelters after dogs and cats, primarily due to owners being unfamiliar with the needed handling, restraint, and nutritional needs of rabbits and all other small mammals.
- **Rabbits Compared to Hares:**
  - **Neither are Rodents:** Rabbits and hares are lagomorphs, not rodents.
  - **Hares:** Hares are larger, with black ear tips. Hares are born in the open with open eyes, fur over the body, and able to run within minutes.
  - **Rabbits:** Rabbits are born blind, naked, and helpless in dens. Rabbits are kept as pets and common laboratory animals; hares are not.
- **Gender and Age Names:** Male rabbits are called *bucks*. Females are *does*, and young rabbits are *kits* or *bunnies*.

NATURAL BEHAVIOR

- **Domesticated Compared to Wild Rabbits:** European (domesticated) rabbits (*O. cuniculus*) have different behavior than the behavior of the North American eastern cottontail rabbit (*Sylvilagus floridanus*). Cottontail rabbits do not burrow and do not tolerate the presence of other rabbits. The European rabbit, the ancestor of the domesticated rabbit, is a social prey animal that lives in burrows of up to 30 individuals.
- **Crepuscular, Social Foragers:** Domestic rabbits like to explore and forage for food, interact with other members of their group, and huddle together when resting. Self-grooming and mutual grooming of others is a frequent activity, and failure to groom can be a sign of disease. They are herbivorous, crepuscular, and nocturnal and like to burrow in soft, sandy dirt.
- **Vulnerable Young:** Rabbits are born without hair and with their eyes closed. Immediate acceptance and care from the mother is essential to survival. Females can be aggressive if their young are perceived to be in danger, but adulteration of the doe's pheromones on kits by handling kits without gloves can lead to the mother's rejection of her babies.
- **Fragile Bones:** Adult size ranges from 2 lbs. to more than 15 lbs. Their bones are fragile compared to other animals of the same size.
- **Gnawing:** Their teeth grow continuously and are normally worn down if allowed to gnaw abrasive food or objects.
- **Coprophagic:** Rabbits are coprophagic and eat cecotrophs directly from their anus about 3 to 8 hours after eating.
- **Defense:** They may thump a rear foot if agitated and may spray urine. They may bite if their head is threatened.
- **Territorial:** Sexually mature rabbits are quite territorial. They assess and claim their territory and possessions by odor. They have glands on their chin and in their perineum, which they use to rub on possessions.
- **Stress from Predators:** Rabbits are prey for many predators, such as dogs, cats, coyotes, ferrets, large birds, and snakes and are therefore naturally fearful of all predators.
- **Prolific Breeders:**
  - **Young Rabbits:** Young rabbits should be separated by gender at 3 months to prevent early matings. In males, the testicles are the most obvious gender-determining structure.

- **Adult Males:** Sexually intact male rabbits can be territorially aggressive and will vocalize (growl, grunt), charge, and claw with their front feet, particularly if threatened by a child, small dog, or cat. Males will need to be housed individually if not neutered.
- **Benefit of Neutering:** Neutered male rabbits, called *lapins*, are more interactive and easy to handle and therefore are better pets for children. Neutered males also are less likely to attempt to mark territory with urine and feces.

APPROACHING AND CATCHING

- **Technique for Catching and Holding** (*Procedural Steps 5.5*):

Procedural Steps 5.5	Catching and Holding Rabbits
1.	Handlers should grasp the skin behind the rabbit's neck while the other hand scoops up the rump (Figure 5.14).
2.	The rabbit should be turned so that its head is tucked under the handler's arm while he maintains a grasp on the neck and supports the hindquarters. This is called the *football hold* (Figure 5.15). Holding the rabbit's body against the handler's chest is the preferred position. Continuing to hold the rabbit by a scruff hold and the other hand supporting the hindquarters without the security of being pressed against the chest is much more stressful for rabbits.
3.	Some rabbits will bite, so care must be taken to avoid putting fingers near their mouth.
4.	Heavy gloves should be worn for protection from scratches if trying to separate fighting rabbits.

Figure 5.14 Rabbits should be scruffed with one hand and their hindquarters supported with the other hand.

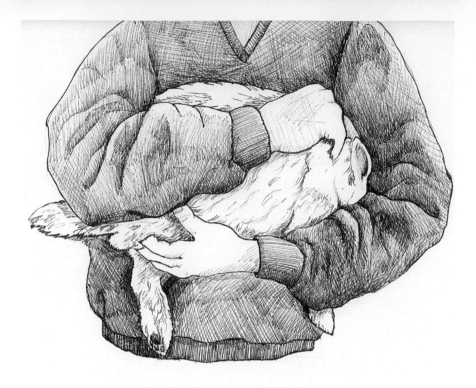

Figure 5.15   The football hold on a rabbit.

- **Avoid Head Restraint:** No effort should be made to restrain or pet the head. Rabbits will strongly resist manipulation of their head, and attempted restraint of the head could lead to a broken neck. Chemical restraint is necessary for examining or treating the head or neck.

HANDLING FOR ROUTINE CARE AND MANAGEMENT

- **Risk of Fractures:** Rabbits will try to twist and kick when resisting restraint, which can cause back injuries, including fractures. Rabbits have thin, light bones and very powerful hind legs. If they kick with suspended hind legs, they can fracture their spine or otherwise damage their spinal cord.
- **Risk of Being Scratched:** In addition, they also have sharp claws that can injure the handler if allowed to kick during handling.
- **Fragile Ear Flaps:** The ear flaps of rabbits are important to their hearing and contribute to heat dissipation. They are delicate structures that should never be used for restraint.
- **Scruffing:** A safe means of restraint is to gently scruff the skin on the back of the neck. If using a scruff hold, the hind legs must be supported and restrained. Lifting a rabbit by a scruff hold alone is likely to result in the rabbit fracturing its back by kicking.
- **Moving in a Crate:** The safest way to move a rabbit is in a travel crate. The crate should be supported underneath with both of the handler's arms, not carried by a handle on top of the crate.
- **Sternal Restraint Method:** Most routine procedures can be performed on rabbits while they are in sternal restraint on a table (***Procedural Steps 5.6***).

Procedural Steps 5.6	Sternal Restraint of Rabbits
1.	A nonslip mat should be placed on the table; otherwise, rabbits may struggle, kick frantically, and fracture bones.
2.	The handler should keep at least one hand on the rabbit at all times.
3.	If holding with one hand, the rabbit's rump should be pushed against the handler's abdomen while one hand presses down on the top of its shoulders.
4.	If restraining with two hands, one hand presses down on the shoulders and the other on the rump.
5.	Toenails can be trimmed with the rabbit in sternal position by lifting one foot at a time.
6.	Addition restraint can be applied with towel wraps or cat bags, as used with cats.

- **Restraint for Ventral Exams:** Examination or treatment of the ventral aspects of the body can be performed by grasping the rabbit's front legs with one hand, turning the rabbit over, and supporting the hindquarters with the other hand. The rabbit's body should then be in a *C* shape.
- **Handling Baby Rabbits:** Special care is required in handling baby rabbits. Handlers should wear plastic gloves and rub the babies with nest bedding when returning them to the nest to keep human odor off the babies.
- **Exit from and Entry to Cages:**
  - **Removal from Cages:** Rabbits should be removed from cages rump first to prevent feet from getting caught in a wire mesh floor.
  - **Placement into Cages:** To place a rabbit into a cage or box, it should go in rump first facing a side wall or facing outward. This prevents it from kicking back and spraying litter out of the box and scratching the handler's arms. The handler should ensure that the rabbit's legs are resting on the surface and ready to support its weight before releasing by pressing the rabbit down and then releasing with both hands at the same time.

FERRETS

European ferrets (*Mustela putorius furo*) evolved from domestication of the European polecat (*M. putorius*). Ferrets have been domesticated since the days of the ancient Roman Empire (300 BC), when they were used to hunt rodents that endangered Roman grain stores.

- **Domesticated Ferrets' Origin:** The black-footed ferret (*Mustela nigripes*), which is indigenous to North America, is a different species and not domesticated. Domesticated European ferrets have been in North America for 300 years.
- **Biopollution Risk:** They are illegal in some states (California and Hawaii), territories (Puerto Rico), and cities (New York City and Washington, D.C.) due to the concern that they would become prolific and prey on indigenous wildlife if the ferrets became feral. This biopollution has occurred in the Shetland Islands and New Zealand.
- **Hearing Deficits:** Domestic ferrets were selectively bred for white haircoats during the Middle Ages so that they could be more easily located. However, ferrets with predominately white haircoats often have Waardenburg syndrome, an inherited trait causing a broadened skull and partial or total deafness.
- **Susceptibility to Human Influenza:** Ferrets are also an animal model for research on human influenza.
- **Gender and Age Names:** Male ferrets are referred to as *hobs*, females as *jills*, spayed females as *sprites*, castrated males as *gibs*, and vasectomized males as *hoblets*. Immature ferrets are called *kits*.

NATURAL BEHAVIOR

- **Cat Size:** Ferrets are approximately the same size as domestic cats, with a longer body and shorter legs. Males are substantially larger than females.
- **Little Thief:** They will hide (*ferret away*) food or favorite toys in an area in their territory that seems the most inaccessible to other animals. Ferret is Latin (*furittus*) for *little thief.*
- **Special Senses:** Ferrets are nearsighted and depend more on detection of odors and their hearing to sense changes within their environment.
    - **Smell:**
        - **Anal Sacs:** Ferret anal sacs are used like scent glands in skunks.
        - **Male Odor:** Most body odor comes from sebaceous glands that are stimulated by male hormones, androgens. Adult males use perianal scent gland secretions, body oils, and sometimes urine and feces to mark their territory and possessions. They will also groom themselves with their urine to attract jills.
        - **Scouting for Food:** They search their surroundings by sniffing the ground and often sneeze. Food odors are important, with olfactory imprinting in young ferrets. Food preferences are developed during their socialization period, between 60 and 90 days of age.
    - **Vision:** Their vision adapts slowly to sudden bright light or darkness. Their pupils are horizontal, in contrast to the vertical pupils of cats. Horizontal pupils may aid in seeing prey (rabbits) with hopping gaits, while vertical pupils may aid tracking of prey (mice) with flat horizontal movements.
    - **Hearing:** The sound range best heard by ferrets is high frequency, 8 to more than 16 kHz, which is the vocal range of their prey.
- **Wary of New Members:** Unlike their more solitary wild cousins, domesticated ferrets like living in groups with established familiarity. A group of domesticated ferrets is called a *business*. New members to the group must be introduced slowly and carefully, because ferrets are territorial.
- **Typical Activities:**
    - **Crepuscular:** Ferrets are crepuscular, most active at dawn and dusk, although they can become imprinted with more diurnal activity during their critical socialization period with humans (4 to 10 weeks of age).
    - **Hunting Play:** They like to chase bouncing hard rubber balls or Ping-Pong balls. Balls should be hard enough to prevent the ferret from eating pieces of them and large enough not to be swallowed whole to prevent developing an intestinal obstruction.
    - **Exploration:** They hunt prey in burrows and are fearless, with short attention spans. They are also extremely curious and will explore every aspect of their environment, especially holes, ducts, rugs, blankets, and tunnels.
    - **Fearless Aggression:** Their fearless aggression is most evident in 3 to 4-month-old males when they play bite each other to establish their group hierarchy and practice their predator skills. Females are more independent and more likely to aggressively bite than males.
    - **Dances:** If excited and happy and wanting to play, ferrets will perform the *dance of joy*, jumping in differing directions in a whimsical manner like that of a baby goat, bumping carelessly into objects. A similar excitement *war dance* will occur with the tail hair fluffed out.
- **Long Periods of Sleep:** When awake, ferrets are boundlessly energetic, but it is normal for them to sleep 12 to 16 hours per day. They like to sleep in enclosed areas or piled with group members.

- **Vocalizations:**
  - **Dook:** Vocalizations include the *dook* (also called chuckling) to express excitement.
  - **Hiss:** If angered or frustrated, they may make a hissing sound, arch their back, and fluff out hair on their tail.
  - **Scream:** If endangered, they will scream.
  - **Bark, Chirp, Squeak:** Barking, chirping, or squeaking is used when a ferret is frightened and defensive.

APPROACHING AND CATCHING

- **Food Lures:** Approaching a ferret is similar to the method of approaching cats. Pastes of food treats on a tongue depressor will attract and distract a ferret for a short time during an acclimation period to a handler. They particularly like fish oil–flavored pastes.
- **Initial Approach:**
  - **Eliminate Hiding Areas for Escaped Ferrets:** If ferrets are allowed to escape during handling, they will seek safety in very small spaces. Therefore, all possible exits, cabinets, drawers, etc. must be blocked prior to beginning a handling session. Ferrets should wear a collar with a bell to help locate them if they escape.
  - **Curled Fingers and Towels:** When approaching a ferret for the first time, the handler should keep his fingers curled when reaching toward it. If unsure of the ferret's disposition, the handler can use a thick towel to cover the ferret and block its vision prior to capture.
  - **Scruffing Resistant Ferrets:** To gain control of a resistant ferret, a handler should scruff the ferret and lift it off its feet. This usually has a calming effect and will cause them to yawn. After the skin on the back of the neck and shoulders is scruffed, the second hand is placed under the rump to support the weight of the body during restraint.

Figure 5.16   Pet ferrets can usually be handled in the same manner as a cat.

- **Picking Up:** Gentle ferrets can be captured and picked up like a cat (Figure 5.16). They often like to hide in a large pocket or bag when carried.
- **Risk of Bites:**
  - **Predatory Bites:**
    - **Keep Ferrets Away from Face:** Domesticated ferrets have retained the predatory characteristics of their wild ancestors: constant searching for prey and aggressive play, including play biting. For this reason, a ferret's face should never be held near a handler's face. Ferrets do not bare their teeth before a bite.
    - **Play Bites:** Ferrets practice their fearless aggression in play fighting with each other. During play fighting, they may bite, hold on, and shake their head right and left. Ferrets, particularly younger ones, may bite a handler as an expression of play aggression.
  - **Fear Bites:** Ferrets are generally docile and receptive to gentle handling used on dogs and cats, but if scared, they can inflict severe bites out of fear. If bitten, a handler should not put a ferret down immediately after a play bite, or it will develop a habit of biting to be released. Due to the risk of bites to the face or fingers, babies and other young children should not handle or be left alone with ferrets.
  - **Startle, Territorial or Food Aggression, and Maternal Bites:** Congenital deafness occurs in some ferrets with white in their haircoat. Deaf ferrets are more easily startled than normal ferrets. Odors on the handlers hand, such as hand lotion or tobacco, may also stimulate a ferret to nip.
  - **Preventing Bites:** Controlling nips and bites should include frequent quiet handling and positive reinforcement of good behavior, beginning during their juvenile period. Removing kits from their mother too early and a lack of maternal discipline can result in a lack of respect for other animals and rougher play aggression. Reactions to bites or nips should be to briefly place the ferret in a special cage or crate (not its normal sleeping cage or crate), without toys or food for a time-out, deprived of attention or reward.

HANDLING FOR ROUTINE CARE AND MANAGEMENT

- **Food Distraction:** Distraction with food treats is sufficient for minor examinations.
- **Picking Ferrets Up:** Greater restraint than simply picking a ferret up and carrying it involves grasping the ferret firmly with one hand around the shoulders and neck. Ferrets are well muscled and require a firm grip. The handler's thumb and index finger should be positioned beneath the jaw to prevent it turning its head to bite.
- **Towel Wraps:** Towel wraps, *kitty burritos*, provide greater restraint when needed.
- **Scruffing:** If even greater restraint is needed or respiratory problems exist, uncooperative ferrets can be scruffed in the same manner as cats.
  - **Method:** Scruffing is performed by firmly grasping the skin on the neck while the hind legs are held to mildly stretch the ferret out.
  - **Precautions:**
    - **Prevent Falls:** Precautions need to be taken so the ferret cannot fall. They should never be scruffed and held above the floor, since they may escape the restraint grip and fall. They should be held over a table with a padded surface.
    - **Avoid Excessive Force:** Ferrets have muscular necks, unlike other small tame animals. Although they must be held tighter than other small animals, excessive force will cause greater struggling and should be avoided.
- **Trimming Nails:** Nail trimming is needed every 2 weeks. This can normally be done while gently holding the ferret in a handler's arms or lap if the ferret was gently handled and desensitized for nail trimming beginning as a juvenile. They have nonretractable claws that should not be removed but require frequent trimming.

- **Avoid Isolation of Individuals:** With few exceptions, small mammals should be moved and housed with group members to reduce stress.
  - **Preferred Companions:** Although any other member of an established group may be helpful, observant handlers may notice certain members may have preferred friends within the group. Including a close friend with the individual needing medical procedures is best.
  - **Facilitated Return to the Larger Group:** Having a companion or companions along with a small mammal can also facilitate the return of all to the larger group.
- **Special Restraint Devices:** Since most small mammals have been used in medical research laboratories, special restraint equipment has been developed for them.
  - **Mice and Rat Tubes (Also Called Tunnels):**
    - **Commercial Tubes:** Plexiglass restraint tubes are commonly used in research to contain and restrain mice and rats. The tubes have a variety of access ports for different procedures. They also have a slot in the top so that a mouse can be restrained by the base of the tail and dragged backwards into the tube.
    - **Syringe Case Tubes:** Restraint tubes, similar to commercial tubes, can be created using the plastic casing for large hypodermic syringes with breathing holes cut into the blind end (Figure 5.17).
  - **Rat Cones:** Rats can also be restrained for medical procedures using a sheet of transparent plastic rolled into a cone or a cake-decorating cone bag. The rat is placed in headfirst and can be held in the cone by one hand (Figure 5.18).

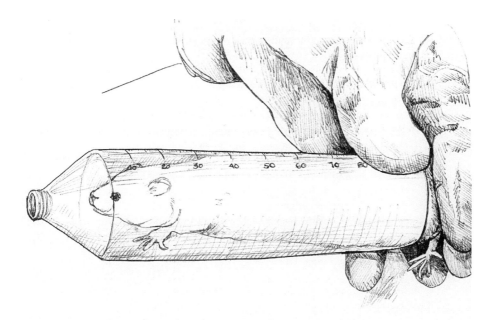

Figure 5.17   Large injection syringes can be modified and used as small rodent restraint chambers.

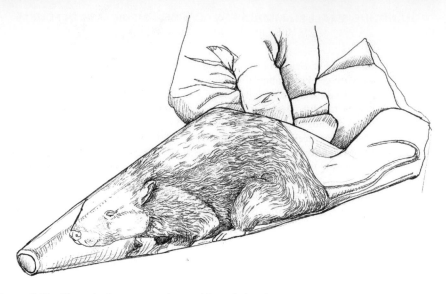

Figure 5.18 Clear plastic cones can be used to restrain rats.

- **Rabbits:**
  - **Restraint Tubes:** Plexiglass restraint boxes, similar to rodent restraint tubes, are available for rabbits.
  - **Towels and Bags:** Rabbits can also be restrained by wrapping in a towel or being placed in a cat bag.
  - **Tables:** Rabbits are very uncomfortable on a slick table, which could lead to violent kicking and possible back injury. Therefore, a mat or rug should be placed on rabbit handling tables. Pressing down on their shoulders while they lie sternal on a table will usually provide sufficient restraint for most procedures.

INJECTIONS AND VENIPUNCTURES

- **Essential Immobilization:**
  - **Area of Transcutaneous Puncture:** Insertion of transcutaneous needles for injection or aspiration in a small mammal carries the risk of slashing tissue beneath the skin, including damage to nerves and blood vessels, and breaking hypodermic needles off in its body.
  - **The Head and Feet:** The area in which the needle is to be inserted must be immobilized, and the animal's mouth and feet should be restrained from interfering with the procedure, especially venipunctures.
  - **Reasonable Comfort for the Animal:** The method of restraint should be comfortable, i.e., no squeezing when unnecessary, but the restraint should allow firm restraint if struggling occurs.
- **Access to Veins**
  - **Purpose:** Venipuncture is necessary to withdraw blood for various analyses or to inject intravenous solutions.
  - **Species Variations:**
    - **Mice, Rats, and Gerbils:** Venipuncture is best done from the lateral vein of the tail while the rodent is in a restraining tube. Lateral saphenous veins are a second choice. The jugular vein may be used in anesthetized rats.

- **Hamsters and Guinea Pigs:** Since hamsters and guinea pigs do not have tails, the lateral saphenous vein is used for venipuncture. Other veins usually require chemical restraint, such as the cephalic veins and ear veins in guinea pigs.
- **Chinchillas:** The lateral saphenous vein is most accessible in chinchillas, but only very small blood samples can be taken from the cephalic, lateral saphenous, or ventral tail veins. The jugular vein is the best site for larger volumes of blood, but anesthesia is needed.
- **Degus:** Venipuncture in degus generally requires chemical restraint. The cephalic and lateral saphenous veins are accessible, but degus become stressed from restraint to perform venipuncture of them. Access to their jugular veins requires anesthesia. Venipuncture of their lateral tail vein is possible, but there is risk of degloving the tail.
- **Hedgehogs:** Venous access in hedgehogs is via the jugular vein during anesthesia.
- **Rabbits:** The lateral saphenous veins are the most accessible in rabbits, but marginal and central ear veins can be used. Required restraint for each is being held in an assistant's arms using the football hold.
- **Ferrets:** The jugular, cephalic, and medial saphenous veins are accessible in ferrets with restraint methods used on cats.

- **Injections**
  - **Subcutaneous**
    - **Restricted to Aqueous Solutions:** Subcutaneous (SC) injections are the safest route for injections, but some injectable materials must not be given SC, such as all oil-based medications.
    - **Method:** Subcutaneous injections can be given in the loose skin of the neck in all small mammals that can be scruffed for handling. The loose skin of the flank is also a possible SC injection site in rats, chinchillas, and ferrets. Fur slip may occur in chinchillas if the skin is lifted (tented) to administer a SC injection.
  - **Intramuscular**
    - **Desirability Is Low:** Intramuscular (IM) injections are undesirable in small mammals due to the risk of muscle, vascular, or nerve damage.

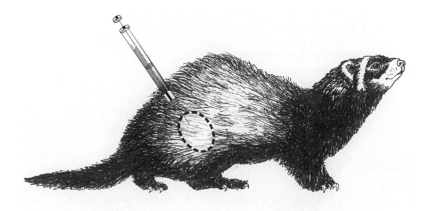

Figure 5.19 IM injection site in a ferret.

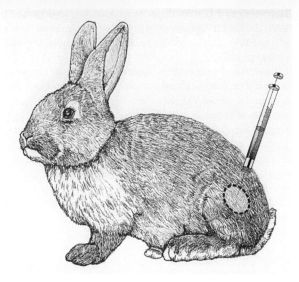

Figure 5.20   IM injection site in a rabbit.

- **Quadriceps Method:** In situations where IM injections are unavoidable, the front muscles of the thigh (quadriceps) are preferable (Figs. 5.19 and 5.20). The leg receiving the injection should be immobilized by the person administering the IM injection with the muscles gently pinched and feeling the location of the femur to help avoid hitting the bone. An assistant handler is needed to provide routine head and body restraint of the animal.
- **Epaxial Method:** The epaxial lumbar muscles are occasionally used for IM injections in chinchillas, rabbits, and ferrets.
- **Intraperitoneal**
  - **Indications:** Intraperitoneal injections (IP) are given into the belly cavity to administer fluids to sick rodents or administer substances for research.
  - **Small Rodents Method:** Small rodents are given IP injections while scruffing the rodent and holding the tail with the little finger. Mice, gerbils, and hamsters are injected into the right caudal quadrant.
  - **Rats Method:** IP injections in rats are administered by a veterinarian or technician while the rat is restrained by a handler who uses one hand to grasp the rat's shoulders and place fingers under the jaw. The handler's other hand supports the rat's body and restrains the hind legs. Rats are injected into the left caudal quadrant.
- **Intraosseous**
  - **Indication:** Intraosseous (into bone marrow) injections are used to deliver fluids to the bloodstream when venipuncture is impossible.
  - **Feasible Species:** Intraosseous injections are not feasible in small rodents but may be performed in large rats, chinchillas, guinea pigs, rabbits, and ferrets using the femur bone.

ADMINISTRATION OF ORAL MEDICATIONS

- **Mixing in Water or Food:**
  - **Drinking Water:** Mixing medications in drinking water is a very unreliable means of medicating small mammals. Undermedication and overmedication can result, and if drinking is completely avoided, the animal will become dehydrated.

- **Food:** Mixing powdered or liquid medication in small amounts of highly palatable food that can only be consumed by an individual animal may suffice if mixing the medication does not adversely affect the medication's actions and the mixture is readily consumed. Possible foods for mixtures can include bread, banana, peanut butter, raisin paste, or fruit-based baby food.
- **Direct Oral Administration Method:**
  - **Restraint:** If oral administration of liquid or soft, pasty medication using restraint is needed, the animal should be held vertically, head up, with the head movement minimized by scruffing or towel or cloth wraps around the neck.
  - **Means of Administration:** Small rodents may be medicated with a gavage needle (curved and blunted) and syringe. Larger small mammals may be administered liquids in the corner of the mouth with an eyedropper.

NOTE

Additional recommended readings on other small mammal handling are available in references on multiple species of small animals provided in the Appendix.

SELECTED FURTHER READING

1. Ballard B, Cheek R. Exotic Animal Medicine for the Veterinary Technician, 3rd ed. Wiley-Blackwell, Ames, IA, 2017.

2. Bays TB, Lightfoot T, Mayer J. Exotic Pet Behavior. Saunders, St. Louis, MO, 2006.

3. Bennett, B. Storey's Guide to Raising Rabbits, 4th ed. Storey Publishing, North Adams, MA, 2009.

4. Campbell KL, Campbell JR. Companion Animals, 2nd ed. Pearson Education, Inc. Upper Saddle River, NJ, 2009.

5. Cloutier S, LaFollette MR, Gaskill BN, et al. Tickling, a technique for inducing positive affect when handling rats. J Visual Exp 2018;135:5–8.

6. Doerning CM, Thurston SE, Villano JS, et al. Assessment of mouse handling techniques during cage changing. J Am Assoc Lab Anim Sci 2019;58:767–773.

7. Gouveia K, Hurst JL. Improving the practicality of using non-aversive handling methods to reduce background stress and anxiety in laboratory mice. Sci Rep 2019;9:20305. doi:10.103/s41598-019-56860-7

8. Judah V, Nuttall K. Exotic Animal Care and Management, 2nd ed. Delmar, Cengage Learning, Albany, NY, 2016.

9. Oxley JA, Ellis CF, McBride EA, et al. A survey of rabbit handling methods within the United Kingdom and the Republic of Ireland. J Appl Anim Welfare Sci 2019;22:207–218.

6

COMPANION BIRDS

DOI: 10.1201/9781003110927-6

Table 6.1	Common Caged Companion Birds
•	Passerines—perching birds
•	Psittacines—hooked-beak parrots and parrot-like birds

Table 6.2	Common Falconry Raptors
•	Accipitrines—hawks and eagles
•	Falcons—falcons and kestrels

Birds have been kept in cages for their beauty and companionship for more than 4,000 years.

There are about 9,000 known species of birds. Caged companion birds are from four orders, and birds of prey (raptors) are from two orders. Major divisions of companion birds are perching birds and hooked-beak birds, while most tamed (falconry) raptors are hawks and eagles or falcons and kestrels (Tables 6.1 and 6.2).

TYPES OF CAGED COMPANION BIRDS AND BIRDS OF PREY

CAGED COMPANION BIRDS

- *Passeriformes*: The largest order is *Passeriformes*, containing the perching birds (canaries, finches, mynahs). Passerines have three front toes and one back toe on each foot. They are also known as songbirds for their ability to vocalize melodies.
- *Psittaciformes*: The order *Psittaciformes* contains the most popular companion birds. Psittacines have two toes in front and two back toes on each foot. Their beaks are hooked, which they use for climbing, breaking nuts, and defense.
    - **Small Psittacines:** Small-sized psittacines include budgerigars, love birds, lories and lorikeets, small conures, Caiques, Pionus, Poicephalus, and cockatiels.
    - **Large Psittacines:** Large-sized psittacines are cockatoos, Amazons, African Grey, large conures, electus, and macaws.
- *Piciformes*: Toucans are the only common companion birds in the order *Piciformes*. Toucans have large, pointed bills that can be as long as their body.
- *Columbiformes*: Pigeons and doves are in the order *Columbiformes*. This order is characterized by a small head and beak, large wings, and a bobbing movement of their head. They have excellent flying ability.

BIRDS OF PREY

- *Accipitriformes:* Hawks and eagles used in falconry are members of the order *Accipitriformes*. They have broad or short wings capable of sharp turns and gliding attacks. They attack primarily with their talons.
- *Falconiformes:* Falcons and kestrels used in falconry belong to the order *Falconiformes*. They have long, pointed wings and are capable of great speed and open plain hunting. They attack with their talons and kill with their beaks.

NATURAL BEHAVIOR OF COMPANION BIRDS

Flock Behavior: Most birds are social animals that preferably live in groups (flocks). Flocks provide added protection, scouting for food sources, and mutual grooming in areas of the body not reachable to groom unassisted.

- **Small Nomadic Birds:** Small nomadic species, such as budgerigars, congregate in large groups for protection from predators.
- **Large Territorial Birds:** Large territorial birds, such as South American parrots, pair bond more strongly and strive for dominance more intensely than small nomadic birds.

CHARACTERISTICS OF COMMON COMPANION BIRDS

- **Characteristics of Orders**
  - **Psittacines:** Cockatiels and budgerigars (parakeets, *Melopsittacus undulates*) are the easiest birds to manage for new bird owners. Both are very social and need frequent interaction with other birds or humans. Budgerigars are smaller and less expensive to own, but they are more flighty and willing to bite when irritated.
  - **Passerines:** Popular passerines are canaries and finches. Canaries prefer to live alone, while finches prefer to live in small groups. Neither canaries nor finches tolerate handling well.
- **Special Senses:** Vision and taste are birds' predominant senses.
- **Vocalizations**
  - **Purposes:** Bird vocalization can be very complex. Vocalization helps coordinate activities such as foraging for food and announcing time to roost and aids in locating mates, establishing territories, and alerting to danger.
  - **Social Acceptance:** Birds that make loud noises, such as screams, are species that will mingle with other avian species in the wild. Birds that do not mingle with other avian species are quieter.
- **Grooming Behavior**
  - **Purposes:** Birds clean and align their feathers by preening, using their mouth to stroke their feathers. They also coat their feathers with an oil from their uropygial gland near the tail while preening. The oil helps waterproof their feathers.
  - **Occurrence:** Preening occurs after bathing and eating.
  - **Allopreening:** Social birds may allopreen, i.e., preen each other.
- **Interactions with Other Birds**
  - **Common Interactions:** Other than mutual grooming, birds do not normally use physical force for interactions with each other. Communications, including dominant aggression, involve vocalizations, posturing, blocking access to resources, and positioning within the immediate surroundings.
  - **Fearful Actions:** Apprehension is often indicated by an open beak while leaning away from a threat from another bird or handler. Fighting is reserved primarily for territorial disputes.
  - **Play Interactions:** Play activities build combat and mating skills and assist in determination of social rank.
- **Respiration in Birds**
  - **Difference from Mammalian Respiration:** Birds do not possess a diaphragm. Their lungs are always filled with air. Air sacs are able to move air in and out. Some of the air from air sacs is delivered to the bones, providing some distributed warmth in cold weather, dissipating heat in warm weather, and adding buoyancy when in water for water birds.
  - **Indication of Heat Stress:** Heat stress causes panting as a last resort and, in some species, rapid fluttering of the throat.
  - **Indications of Cold Exposure:** Exposure to cold weather leads to fluffing of feathers to trap insulating pockets of air and sitting on their feet to keep their feet warm.
- **Body Language:** Bird mannerisms include alternating pupil dilation and constriction and flaring tail feathers when excited. Wings are spread when acting secure. Puffing out the feathers momentarily or wagging the tail signals a greeting.

- **Typical Activities:** The major activities are being on alert for predators and foraging for food. Although both can be stressful, these activities are important in maintaining normal mental health and behavior.

SAFETY FIRST

Many companion birds may enjoy interactions with humans, but none enjoy being restrained. Always reassess the need for whether a bird must be handled and restrained before subjecting it to those stresses.

HANDLER SAFETY

- **Nondefensive Use of Beaks:**
  - **Acceptable Beak Use:** Companion birds use their beaks to balance going from perch to perch. Most only aggressively bite as a last resort when frightened. They also use their beaks and tongues to investigate their surroundings by touch and taste.
  - **Unacceptable Beak Use:** Large psittacines may make biting a game if they can evoke a reaction from a bitten handler.
- **Means of Bird Defense:**
  - **Species Variations:**
    - **Passerines:** Small, straight-billed perching birds (finches, canaries) resent being handled and will defensively stab or bite.
    - **Psittacines:** Parakeets, parrots, and other psittacines may also bite. Since large parrots can crack walnuts with their beaks, they can just as easily break a finger.
    - **Raptors:** Birds of prey primarily use their talons to attack.
    - **Columbiformes:** Pigeons and doves are not aggressive and pose no physical threat to handlers.
  - **Proper Handling Attire:**
    - **Urofecal Soiling Protection:** When handling any bird, a handler must expect to be defecated on and should wear appropriate outer clothing.
    - **Hearing Protection:** Ear protection is advisable if handling a large psittacine screamer.
    - **Avoidance of Bright and Shiny Objects:** The sense of sight in birds is excellent, and birds, like most animals, are very inquisitive. Handlers who wear bright colors or shiny jewelry invite being pecked when handling birds.
- **Aggression:**
  - **Striving for Elevated Positions:**
    - **Establishing Dominance Over Handlers:** Birds demonstrate their dominance over other birds by assuming a higher perch position. Allowing a bird's head to be above the handler's eye level by the bird resting on the handler's head or shoulders gives the bird the impression it is dominant to the handler. A handler should not hold them higher than the handler's mid-chest level.
    - **Greater Risk to Handlers:** Furthermore, allowing a parrot to perch on a shoulder positions them in a way that the handler cannot control them well and invites bites to the handler's ears, neck, lips, and possibly the eyes.
    - **Risk to Birds:** The lack of control from shoulder perching can also increase the risk of serious injury to the bird, if it becomes suddenly startled.
  - **Reprimands:** Attempts to bite should be reprimanded by either being startled, i.e., suddenly dropping the hand the bird is perched on a short distance, or human attention should be taken away by isolating the bird from human attention for a short period.

BIRD SAFETY

- **Reduce Stress of Handling with Early Socialization**
  - **Tolerance to Handling:** Hooked-beak birds such as parrots are generally more tolerant of being handled. They can be socialized with humans and may bond with a human family member if socialized while young.
  - **Hand-Raising Young Birds:** Birds that are hand-raised are imprinted with humans and require more human attention for a feeling of security.
- **Brittle Bone Danger:** Most birds will resist being handled and endanger themselves trying to escape. For most species of birds, restraint of the wings is the first objective for handlers. Bird bones are very light and break easily, particularly wing or leg bones.
- **Respiratory Hazards**
  - **Impairment of Respiratory Movements**
    - **Possible Asphyxiation and Shock:** Handlers must remain mindful that it is easy to restrict a bird's respiratory movements by holding them too tightly, which can cause unnecessary struggling and lead to shock.
    - **Restrain Neck:** The sternum's movement must not be restricted, or they cannot breathe. Their tracheal rings are complete and relatively resistant to collapse when birds are held by the neck.
    - **Avoid Pressure on Chest:** Birds do not have a complete diaphragm, and the lungs are associated with the chest wall. Slight compression around the chest during restraint can eliminate their ability to breathe.
  - **Risk of Pneumonia:** Birds are very susceptible to pneumonia caused by chilling or exposure to drafts; therefore, a bird should never be placed near an open window or air conditioner vent.
- **Risk of Overheating:** Feathers trap air and provide efficient insulation. However, when birds are handled, the insulation of their feathers can predispose them to becoming overheated. Physical restraints should be used for the shortest period possible to reduce the risk of the bird overheating.
- **Damaged Wing Danger:** Damage to flight feathers may endanger birds that are released for sport such as pigeons and raptors.
- **Danger from Predator Pets:** Birds should never be left unsupervised with dogs, cats, ferrets, reptiles, or children.
- **Risks of Leg Banding:** Banding young birds can put them at risk of constricting the leg during growth or being caught on objects by a loose band.
- **Household Dangers:** Companion birds should not exercise freely in a house. The dangers are numerous (Table 6.3).

Table 6.3 Household Dangers for Companion Birds	
•	Poisonous plants or household pesticides
•	Electrical cords
•	Eating carpet
•	Electric fans
•	Heaters and stoves
•	Entanglement in terrycloth towels
•	Becoming caught in open toilets
•	Inhalation of ammonia from cleansers
•	Predator pets
•	Escape through open doors or windows

KEY ZOONOSES

Apparently healthy captive-bred, caged companion birds pose little risk of transmitting disease to healthy adult handlers who practice conventional personal hygiene. The risks of physical injury are greater than the risks of acquiring an infectious disease (Table 6.4).

SANITARY PRACTICES

When handling more than one bird from different households, proper sanitation is required to prevent the spread of disease from carriers without clinical signs.

- **Sanitary Practices for Handlers** (Table 6.5)
- **Stressing Birds Spreads Disease:** Stressing birds can lower their immunity and should be avoided. Stress can cause shedding of psittacosis and other diseases. Dust from bird enclosures and feathers should be controlled.
- **Sick Bird Precautions:** Special precautions are needed if sick birds are handled (Table 6.6).

Table 6.4 Diseases Transmitted from Healthy-Appearing Caged Companion Birds to Healthy Adult Humans

Disease	Agent	Means of Transmission	Signs and Symptoms in Humans	Frequency in Animals	Risk Group*
Bites, nail, or talon injuries	——	Direct injury	Bite wounds to face, hands, or arms	All birds are capable of inflicting bites or wounds.	3
Salmonellosis	*Salmonella typhimurium*	Direct, fecal-oral	Diarrhea, systemic infections	Common in birds	3
Psittacosis	*Chlamydiophila psittaci*	Direct, fecal-oral and respiratory secretions Indirect from fomites and bird mites	Pneumonia	Fairly common in birds	3
Avian tuberculosis	*Mycobacterium avium*	Direct, fecal-oral, respiratory secretions	Pneumonia	Avian TB is rare in caged birds. Most TB in caged birds is from infected humans.	3
Avian influenza	Type A H5N1	Direct, respiratory secretions	Mild conjunctivitis and respiratory distress to possible death	15-state outbreak in domestic birds in 2014–15	4

*Risk Groups (National Institutes of Health and World Health Organization criteria. Centers for Disease Control and Prevention, Biosafety in Microbiological and Biomedical Laboratories, 5th edition, 2009.)
1: Agent not associated with disease in healthy adult humans.
2: Agent rarely causes serious disease, and preventions or therapy possible.
3: Agent can cause serious or lethal disease, and preventions or therapy possible.
4: Agent can cause serious or lethal disease, and preventions or therapy are not usually available.

Table 6.5	Sanitary Practices for Bird Handlers
•	Handlers should wash their hands and clean and disinfect tabletops and cages used in handling.
•	A handler of birds should wear appropriate dress to protect against skin contamination with feathers and skin scales or fecal droppings.
•	Hands should be kept away from eyes, nose, and mouth when handling birds. Wash hands after handling birds.
•	Handlers should not kiss birds.
•	A face mask should be worn when cleaning cages and floors.
•	After cleaning, cages should be disinfected with 1:100 solution of bleach to water (2 tablespoons/gallon of water).
•	Immunocompromised handlers should not clean cages.

Table 6.6	Precautions for Sick Birds
•	Sick birds should be isolated from apparently normal birds.
•	New group members should be quarantined for at least 2 weeks to reduce the risk of transmitting a disease that new birds could be incubating before introducing to the rest of the group.
•	Birds from different origins should not be confined in the same cage.
•	The origin of birds should be verified to prevent buying illegally imported birds.
•	Legal importation requires a quarantine period and prophylactic treatment for psittacosis.

APPROACHING AND CATCHING

REASONS TO HANDLE BIRDS

- **Socialization:** Companion birds need to be handled for socialization to humans, environmental enrichment, and to determine their body condition.
- **Monitoring Health:** Feathers obstruct the visual assessment of loss of muscle mass or the development of abdominal enlargement. Early detection of diseases requires palpation of the bird's body.

ASSESSMENT OF HANDLING TOLERANCE

- **Danger:** Handling a bird that is overly stressed due to health reasons or fear of being handled can be life-threatening to the bird.
- **Signs of Intolerance:** Handling should be avoided in birds that demonstrate respiratory distress, potential injurious attempts to escape or avoid handling, or extreme depression. The bird should be allowed to acclimate to the surroundings until calm or provided with critical care in a veterinary intensive care unit.

MOVING BIRDS TO BE HANDLED

- **Small Birds and Tame Large Birds:** Small birds and tame large birds should be moved in their cage to prepared areas for handling.
- **Large Birds Unaccustomed to Being Handled:** Large birds, such as adult parrots, that are unaccustomed to being handled should be transported in a crate,

preferably with a top door. Crates without a top door that allows approach to the bird from above its head should be tipped on end so that the front door becomes a top door.

Birds Capable of Normal Flight: Adult birds that have full ability to fly, i.e. their wings have not been clipped, are generally best handled in their carrier or cage.

ENVIRONMENTAL PREPARATION

Before attempting to capture a caged companion bird, the room and the cage should be prepared (*Procedural Steps 6.1*).

Procedural Steps 6.1	Preparation of the Environment for Handling Birds
1.	Light from windows should be blocked with shades or blinds to reduce the chance that if the bird escapes, it will try to fly through the glass.
2.	Heaters and fans should be turned off.
3.	Vents, windows, and doors should be closed.
4.	Bowls, toys, and other objects (so-called *furniture*) in the cage that are not attached to the bars should be removed from the cage (Figure 6.1).

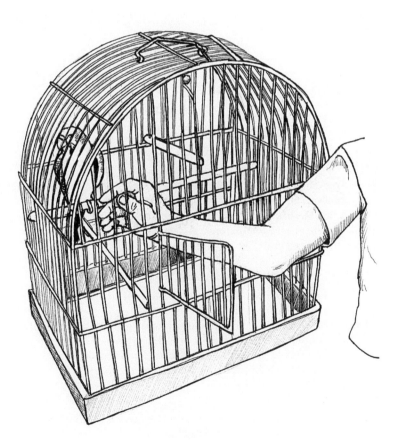

Figure 6.1   All loose *furniture* should be removed from a cage before capturing a caged bird.

Handler Preparation: Handlers should wash their hands before handling each bird for sanitation and to reduce odors of other birds or predators (dog and cat odors).

IMPACT OF SOCIALIZATION

- **Early Socialization is Essential for Low-Stress Handling**
    - **Benefits:** Handling of birds is greatly facilitated if the bird is properly socialized and handled between 4 weeks and 3 months of age.
    - **Procedure:** Young birds should be desensitized to handling with towels, trimming nails, and trimming of flight feathers. Mirrors in cages may reduce bonding with handlers and should be avoided.
    - **Frequency:** Pet birds tolerant of handling (primarily hooked-bill birds) should be handled daily and allowed to exercise outside their cage. At least 10 hours of quiet sleep is important.
    - **Bad Behaviors:** Bad behavior, nibbling fingers, biting, or screaming should be ignored. Bird behavior is not altered by reprimands. Some may be emboldened by getting reactions from handlers when exhibiting bad behaviors.
- **Catching Socialized Birds (*Procedural Steps 6.2*):**

Procedural Steps 6.2 Catching Socialized Companion Birds	
1.	A slow approach and offering a small food treat prior to handling can be helpful in reducing resistance.
2.	The handler should present a horizontal index finger slowly toward the bird's breast, and the bird will step up onto the finger. Birds will step up onto a finger but usually not over or down on a finger.
3.	To ask a bird to step off of a finger, it should be presented to the perch at its breast level.

- **Catching Inadequately Socialized Birds (*Procedural Steps 6.3*):**

Procedural Steps 6.3 Methods of Catching Poorly Socialized Caged Birds	
Bare-Hand Capture	
1.	If a small bird to be captured has not been trained to step up onto a finger, it must be caught barehanded from behind.
2.	After positioning the hand for capture near the slightly opened door of a small bird's cage, an assistant should briefly turn off the room lights to distract the bird and briefly impair its vision to facilitate the final capture.
3.	The bird should be grasped around its neck from behind with an index finger and thumb while cupping the remaining fingers around the wings and chest.
Towel Capture	
1.	Untamed birds in small cages may be better approached using a towel over the handler's capture hand.
2.	The bird is captured from behind with fingers around its neck and cupping the body loosely.
3.	The hand on the body is moved to the feet, and a finger is placed between the feet being held.
4.	The bird's body should be held against the handler's body to restrain one wing while the fourth and fifth fingers of the hand on the head are extended over the other wing for restraint.
5.	The towel is folded back off its head after capture.

Procedural Steps 6.3	Methods of Catching Poorly Socialized Caged Birds
Capture by Net	
1.	In a large cage or aviary, a net may be necessary for capture.
2.	After net capture, the bird's neck is grasped from behind.
3.	The net is carefully removed while retaining control of the bird's neck and holding the wings close to the bird's body without squeezing its chest.

HANDLING FOR ROUTINE CARE AND MANAGEMENT

BASIC EQUIPMENT

Nearly all handling of birds is manual. Restraint tubes, muzzles, squeeze cages, etc. are not used with bird handling. The basic handling equipment for birds is an angular (square or rectangular) cage and towels.

RESTRAINT OF INDIVIDUALS OR PORTIONS OF THEIR BODIES

- **Whole Body**
  - **Bare-Hand Hold (*Procedural Steps 6.4*):**

Procedural Steps 6.4	Whole-Body Restraint of Birds Using a Bare-Hand Hold
1.	A bird's body is restrained by holding the wings against their body and controlling movement of the head.
2.	It is important to not impair the ability of their chests to easily expand, so a handler's fingers should be separated when handling birds to reduce compression on the thorax and ability to breathe.
3.	Care is also needed to avoid damage to their plumage.

- **Holding with a Towel (*Procedural Steps 6.5*):**

Procedural Steps 6.5	Whole-Body Restraint of Birds Using a Towel
1.	Towels can be used, but looped-thread cloth should be avoided, because the loops can catch the bird's nails.
2.	Gloves should not be used except with raptors (hawks and owls).
3.	Capturing the bird should be done by approaching it from behind and placing the towel over the bird's head and grasping around its body and wings.
4.	The cloth over the head is then folded back as if removing a hood.
5.	The restrained bird should be held close to the handler's body to provide a better feeling of security in the bird.
6.	Macaws may protest loudly enough that ear protection is advisable.

- **Restraint Jackets:** Commercial avian straitjackets are available that fold over the wings and wrap around the body with a Velcro closure. The bird requires little to no further restraint after the wrap is applied. However, commercial jackets require two handlers for birds that resist restraint and are applied more slowly than towel wraps.
- **Head:** A bird's head is typically restrained by a hand with the palm behind the head with either an index finger and thumb or an index finger and middle finger positioned on each side of the neck and under the jaw, restraining head movement. This is the same neck-collar hold used on rodents.

- **Wings**
  - **Wing Trimming:**
    - **Prevention of Escape:** The escape of companion birds by flying out opened doors and windows can be reduced by trimming 2 to 1/3 of the ends of their primary flight feathers, which are approximately 4 to 10 feathers on each wing.
    - **Prevention of Injuries:** Clipping flight feathers can also eliminate or reduce the risk of flying into glass of windows, onto hot cooking ranges, or the blades of fans. Aggressive birds may need to be clipped to protect owners, handlers, or other animals.
    - **Trimming versus Pinioning:** Trimming is not pinioning. Pinioning is an amputation of the wing at the carpal joint.
  - **Method:**
    - **Age:** Wing clipping should not be performed until fledglings learn to fly to prevent behavioral disorders associated with a lack of confidence.
    - **Procedure:** Trimming is performed by extending a wing and clipping the ends of the feathers with sharp, unsprung scissors. Birds that have not been trained when young to accept restraint and wing extension may need to be wrapped in a towel by an assistant and to have a wing extended. The regrowth of feathers should be checked 6 to 8 weeks after a trim to determine if retrimming is due.
  - **Disadvantages:**
    - **Predisposed to Some Injuries:** Although wing clipping can be a safety precaution, it can render the bird more vulnerable to other dangers such as being stepped on or injured by other pets.
    - **Modified Flight and Unsafe Landings:** Mild wing clipping or strong wind currents in the outdoors can enable many birds to still fly, but their ability to control a landing may be impaired and result in injury. Clipping of wings should be bilaterally symmetrical, or the bird will be imbalanced when attempting to fly and may injure itself.
    - **Complete Loss of Flight:** Clipping so closely that any ability to fly is lost can cause injury to the sternum (broken keel bone) if the bird attempts to fly or falls from an elevated position. Wing clipping should be done in increments to allow the bird to adapt to the inability to fly and different coordination needed to maintain balance. It should never be severe enough to prevent the bird's ability to glide to the floor.
    - **Blood Loss:** During molting season, new growing feathers have an abundant blood supply and are referred to as blood feathers. Blood feathers should not be trimmed, or significant blood loss may result.
- **Mouth**
  - **Beak Trimming:** Birds that are not given sufficient opportunities to grind down the growth of their beak require trimming of the beak. Use of cuttlebones, concrete perches, or other abrasives in the cage usually eliminates the need to trim beaks.
  - **Methods:**
    - **Restraint:** The mouth is not specifically restrained. The bird's head is restrained while the beak is trimmed.
    - **Rotary Grinder:** A hand-held rotary grinder is often used to achieve a normal-shaped beak. The noise of the running grinder should be introduced at a distance to the bird, and based on signs that the bird becomes desensitized to the noise, the grinder is moved closer until it can be used briefly on the beak. With repeated brief use, the duration of use can be increased.
    - **Emery Board:** An emery board may be used in lieu of a rotary grinder.

- **Legs**
  - **Toenail Trimming:** Toenails can also overgrow if normal opportunities for abrasion of the toenails do not exist. Providing one cement perch in addition to wooden perches will usually eliminate the need to trim nails.
  - **Training for Manual Trimming:** Training for toenail trimming should begin when birds are in their socialization period. Only the tip of the toenail should be clipped or ground down.
  - **Method:**
    - **Restraint:** Restraint of a leg during toenail trims is done by grasping and extending the leg while the body is held with a neck-collar hold and fingers cupped around the wings and chest. Larger birds are held by an assistant while the person trimming the nails extends and holds the leg with the toenails to be trimmed. The same restraint is used to apply identification leg bands, but embedding a microchip into the left pectoral muscle of the bird is preferred by most handlers for identification.
    - **Trimmers:** Depending on the size of the bird, toenails may be trimmed with human nail clippers, dog nail trimmers, or hand-held rotary grinders.

HANDLING VARIATIONS FOR DIFFERENT BIRD SIZES AND TYPES

- **Small-Sized Birds** (budgerigars/parakeets, canaries, finches)
  - **Towel Restraint:** If full-body restraint is needed, a towel can be placed over the handler's hand to mask the hand approaching. However, handling small or medium-sized birds with towels or cloths may cause them to become overheated.

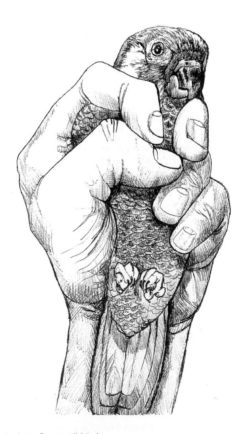

Figure 6.2   One-hand restraint of a small bird.

- **Bare-Hand Restraint:**
  - **Common Hold:** The neck is grasped between an index and middle finger or the index finger and thumb. The chest is held loosely with fingers spread apart to aid in avoiding the restriction of chest movements (Figure 6.2). The bird's feet may be allowed to grasp the handler's little finger.
  - **Alternative Hold:** An alternative grip is to hold the neck between the thumb and middle finger with the index finger on the top of the head.
- **Medium-Sized Birds** (pigeon- to hawk-sized: cockatiels, cockatoos, conures, doves, parrots)
  - **Two-Hand Lateral Restraint:** The bird's body must be grasped with both hands, but respiration cannot be restricted. The wings and chest are held gently on both sides by two hands with fingers separated.
  - **Two-Hand Upper/Lower Restraint:** Alternatively, the head is restrained by using the handler's thumb and index finger on the neck while holding the wing tips (distal remiges) and the legs (tibiotarsal bones) (Figure 6.3). A firmer and more comfortable grip on the legs is with the thumb and middle finger around the legs and the index finger between the tibiotarsal bones.

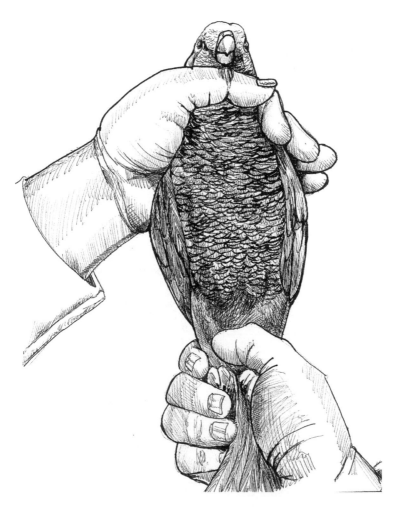

Figure 6.3 Two-hand restraint of a medium-sized bird.

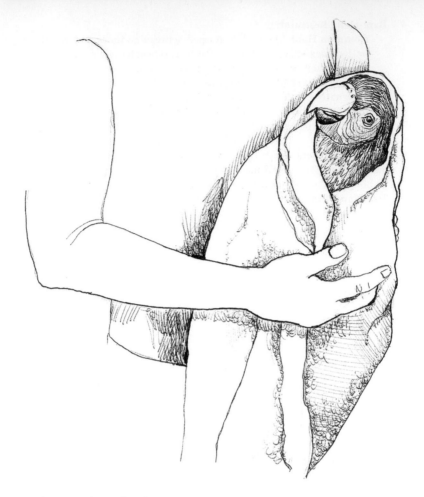

Figure 6.4   Towel restraint of a parrot.

- **Large-Sized Birds** (Amazon, African Grey, macaw, cockatoo parrots)
  - **Risk to Handlers:** Large-sized birds can be dangerous to handlers. For example, macaws can have a wingspan of up to 4 feet and extremely strong jaw strength to inflict severe bites.
  - **Method (*Procedural Steps 6.6*):**

Procedural Steps 6.6	Handling of Large-Sized Caged Birds
1.	Parrots defend themselves primarily with their beaks so their heads must be secured first. Towel restraint should be used if the bird is resistant to being identified. Gloves should not be used.
2.	Approaching slowly from the front will cause less distress in the bird.
3.	A handler wraps the towel over the head and around the wings (Figure 6.4).
4.	The neck is grasped between the thumb and fingers with the tips of the fingers beneath the lower aspect of the jaw (Figure 6.5). At the same time the bird's feet are grasped with the other hand and then the bird is held next to the handler's body.
5.	A nontoxic wooden stick can be offered for the parrot to bite as a distraction, if needed.

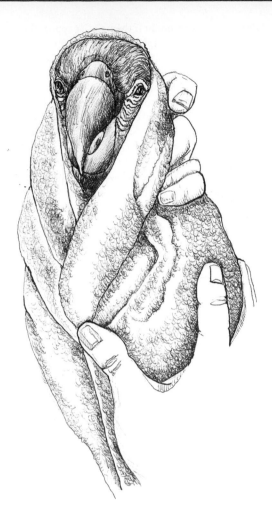

Figure 6.5   Restraint of the head with a towel.

- **Raptors/Birds of Prey** (hawks, kestrels, falcons, eagles)
  IMPORTANT: A State Rehabilitator's license is required to care for and rehabilitate sick or injured wildlife, and a Federal Special-Purpose Rehabilitation Permit is needed in order to care for and rehabilitate migratory birds and endangered or threatened species of wildlife such as raptors. To keep a raptor, a handler must serve a 2-year apprenticeship, pass a written exam, build acceptable facilities, and maintain thorough records of care.
  - **Raptor Restraint Equipment:** Permanently captive raptors usually have jesses (leather straps attached to grommets in leather anklets) for easier leg restraint.
  - **Method:**
    - **Capture Feet First:** Raptors use their talons for primary defense, so their feet must be secured first.
    - **Wear Gloves with Gauntlets:** A handler should never take off leather gloves with gauntlets when handling a raptor.
    - **Adjust Room Lighting:** Dimmed lights in a room for diurnal birds (hawks) and bright lights for nocturnal birds (owls) can create an environment more conducive for quiet handling.

- – **Direction of Approach:** If the handler approaches from behind, the wings, body, and legs are grasped together, and if the handler approaches from the front, the legs are grasped first.
  - – **Shield Handler's Face:** Appropriately thick leather gloves should be used. One gloved hand should be kept between the bird and the handler's face as protection and a distraction while the bird's feet are grasped with the other hand by placing an index finger between the feet.
  - – **Expect Bating:** If excited, raptors will occasionally bate, i.e., attempt to escape, flip over, and hang by the jesses.
- **Wading Shore Birds (herons, cranes, egrets, and pelicans)** (*Procedural Steps 6.7*)

  IMPORTANT: Wading shore birds use their long, pointed beaks for defense. When handled, they pose a great danger to handlers' eyes. Protective eyewear must be worn whenever these birds are handled.

Procedural Steps 6.7	Capture, Restraint, and Release of Wading Shore Birds
Capture and Restraint	
1.	Wading shore birds should be approached from behind.
2.	The handler should reach from behind the bird and grasp the beak from underneath it.
3.	At the same time, the other arm reaches over the bird's body and grasps the feet.
4.	Large shore birds may require two handlers for restraint.
Release	
1.	The bird should be placed in its enclosure positioned facing away from the handler.
2.	Its head is held with an outstretched arm.
3.	The head is released last as the handler's arm is withdrawn, and the enclosure door is quickly closed.

RELEASING IN A CAGE

- **Prepare the Cage:** A bird should be returned to its cage after removing all toys, perches, bowls, and other cage materials.
- **Method** (*Procedural Steps 6.8*):

Procedural Steps 6.8	Releasing a Bird into a Cage
1.	The bird is placed on the cage floor, and the door is closed against the handler's arm that restrained the bird.
2.	The arm should be carefully withdrawn while keeping the door closed on it until the door can be latched.
3.	Finally, while opening the door a minimal degree, cage materials are returned to the cage as quietly as possible.

HANDLING FOR COMMON MEDICAL PROCEDURES

Most handling and restraint of birds can and should be done without tranquilization, sedation, hypnosis, or anesthesia.

INJECTIONS AND VENIPUNCTURE

Insertion of transcutaneous needles for injection or aspiration in birds carries the risk of slashing tissue beneath the skin, including damage to nerves and blood vessels, and breaking hypodermic needles off in its body. The area in which the needle is to be inserted must be immobilized, and the bird's mouth and feet should be restrained from interfering with the procedure, especially venipunctures. The method of restraint should be comfortable, i.e., no squeezing when unnecessary, but the restraint should allow firm restraint if struggling occurs.

- **Venipuncture:** Venipuncture is performed to collect blood for diagnostic analyses or intravenous injections.
    - **Right Jugular Vein in Small Birds:** Blood samples are usually collected from the right jugular vein in small birds. The left jugular is much smaller. The crop should be empty prior to venipuncture to reduce the risk of regurgitation and aspiration.
    - **Leg or Wing Veins in Larger Birds:** The preferred veins in larger birds for venipuncture are the brachial veins underneath the wings near the body, the ulnar veins underneath the wings more distal to the body, or the medial metatarsal veins on the legs.
    - **Restraint Jackets:** Birds can be restrained for venipuncture from the jugular vein or leg veins with a restraint jacket, which has a flexible back with two straps that attach in front of the keel with Velcro.
- **Injections**
    - **Intramuscular:** Intramuscular injections are typically given into the ventral part of a pectoral muscle (Figure 6.6).

Figure 6.6   Intramuscular injection site in birds.

- **Subcutaneous:** Subcutaneous injections are administered into the inguinal skin fold immediately in front of a leg or the axillary area under a wing.
- **Intraosseous Injection:** Intraosseous injections require sedation or anesthesia and are used only for fluid administration. In small birds, intraosseous injections are given into the proximal tibiotarsal bone below the stifle joint. The distal and proximal ulna can also be used in large birds.
- **Intraperitoneal:** Intraperitoneal injections should never be administered to birds, because a needle could rupture the air sacs.

ADMINISTRATION OF ORAL MEDICATIONS

- **Indirect Administration:** Adding medication to food and water is the least stressful method of oral administration, but it is also the most unreliable for delivering the desired dose. Medications may be unstable in water, or they may make the drinking water unpalatable. Direct administration is more reliable for accurate dosing.
- **Direct Administration Methods:**
    - **Syringe or Dropper** (*Procedural Steps 6.9*):

Procedural Steps 6.9	Syringe or Dropper Oral Administration to Birds
1.	To directly deliver oral liquid medication, the handler should put the tip of the syringe or dropper with medication into the corner of the bird's beak and direct the tip toward the other side of the mouth, not toward the throat.
2.	While talking quietly to the bird, the handler should deliver the medication slowly, providing sufficient time for swallowing.
3.	The tip of the syringe or dropper should be maintained inside the mouth at the corner of the beak until done.
4.	A treat should be offered while continuing the restraint until the bird is quiet. Then the bird is released slowly.

   - **Crop Tube** (*Procedural Steps 6.10*):

Procedural Steps 6.10	Crop Tube Oral Administration to Birds
1.	The required distance of the tube to the crop, which is just below the base of the neck, should be determined and the crop tube marked prior to placement in the bird.
2.	When placing a crop tube, the handler elevates the bird's beak to place the feeding tube through the mouth and into the esophagus.
3.	Insertion of the tube should stop when the mark on the tube indicates the end is in the crop.

   - **IMPORTANT—Minimize Handling:** Birds should not be handled or restrained after oral medication is directly administered because of greater risk of inducing regurgitation and aspiration.

SPECIAL EQUIPMENT AND PROCEDURES

Special handling equipment or procedures for birds can include gloves and gauntlets, bags, tubing, lighting, harnesses, and leg and beak bindings.

BAGS
Bags for restraint of birds can include stockings, pillowcases, and other cloth bags.

CARDBOARD TUBING

Toilet paper or paper hand towel cardboard tubes can be placed around appropriate-sized birds for temporary restraint. The purpose is to impair movement of the wings and legs but not compress the breathing movements of the thorax.

LIGHTING

Diurnal birds are quieter if deprived of bright lighting. Diurnal birds of prey (falcons, hawks, eagles) are kept quieter with the use of hoods. Nocturnal birds (owls) are quieter in bright lighting.

HARNESSES

Bird harnesses are used with a leash for exercising birds. A harness has a loop that goes over the bird's head and is connected to a loop that goes around their body behind the wings.

BEAK BINDING

Beak binding can aid in the restraint of straight-beaked birds. Toucans and other large, pointy-beaked birds can be aggressive with their beaks.

- **Elastic Bands:** While wearing a face shield and leather gloves, the handler should grasp the bird's bill. The bill is then taped shut or bound with elastic bands.
- **Tennis Ball:** In addition, a large cork or a tennis ball with a slit cut in it should be pushed onto the point of the beak.

RESTRAINTS FOR RAPTORS

- **Hood:** Hawks are diurnal hunters that depend on their vision. Hooding them quiets their activity, reduces their startle reactions, and allows them to be carried without exciting them.
- **Gloves and Gauntlets:** Gloves and gauntlets are made of leather and are typically used for handling raptors. Gauntlets are heavy gloves that extend beyond the wrist and protect the forearm.
- **Leg Bindings**
  - **Jesse:** A jesse (pronounced *jess*; last *e* is silent) is a leather strap about 8 to 9 inches long that is strapped to each leg.
  - **Leash:** A leash is a leather strap about 3 feet long that can be attached to the jesses.
  - **Bewits:** Bewits are leather strips that tie bells to the feet of raptors.
  - **Creance:** A creance is a long, lightweight line that is tied to the perch on one end and the jesses on the other for teaching a raptor to fly from the perch to the handler's gloved fist.

NOTE

Additional recommended readings on bird handling are available in references on multiple species of small animals provided in the Appendix.

COMPANION BIRD HANDLING REFERENCES AND SUGGESTED READING

1. Ballard B, Cheek R. Exotic Animal Medicine for the Veterinary Technician, 3rd ed. Wiley-Blackwell, Ames, IA, 2017.

2. Bays, TB, Lightfoot T, Mayer J. Exotic Pet Behavior. Saunders, St. Louis, MO, 2006.

3. Campbell KL, Campbell JR. Companion Animals, 2nd ed. Pearson Education, Inc. Upper Saddle River, NJ, 2009.

4. Judah V, Nuttall K. Exotic Animal Care and Management, 2nd ed. Delmar, Cengage Learning, Albany, NY, 2016.

5. Warren DM. Small Animal Care & Management, 4th ed. Delmar, Cengage Learning, Albany, NY, 2015.

7

REPTILES

DOI: 10.1201/9781003110927-7

Table 7.1	Groups of Common Pet Reptiles
•	Chelonians—turtles and tortoises
•	Serpentes—snakes
•	Saurians—lizards

Reptiles are nondomesticated animals with scales or scutes for a body exterior that lay eggs and are cold-blooded. Many reptiles are docile and can be kept as tame pets, although they do not tolerate, nor can they endure, much handling. More than 7 million reptiles are currently kept as pets in the U.S. Major divisions of common pet reptiles are turtles/tortoises, snakes, and lizards (Table 7.1).

TYPES OF PET REPTILES

Chelonians: Chelonians are also referred to as Testudines.

- **Types**
  - **Turtles:** Turtles live most of their lives in or near water. They have webbed feet with long claws. Turtles move faster than tortoises.
  - **Tortoises:** Tortoises are terrestrial and have thick, elephant-like feet with stubby claws.
  - **Terrapins:** Terrapins are semiaquatic, hard-shelled chelonians. The term "terrapin" is ill-defined and is derived from a term used by the early European colonists in North America for edible turtles.
- **Popularity as Pets:** The most popular pet turtles in the U.S. are the red-eared slider (*Trachemys scripta elegans*), eastern box turtle (*Terrapene carolina carolina*), western painted turtle (*Chrysemys picta bellii*), map turtle (*Graptemys geographica*), and wood turtle (*Glyptemys insculpta*).

SQUAMATES

- **Serpentes:** The most popular pet snakes are the ball python (*Python regius*), corn snake (*Pantherophis guttata*), California kingsnake (*Lampropeltis getula californiae*), rosy boa (*Lichanura trivirgata*), and gopher snake (*Pituophis* spp.).
- **Saurians:** Popular pet lizards include the bearded dragon (*Pogona vitticeps*), gold-dust day gecko (*Phelsuma laticauda*), leopard gecko (*Eublepharis macularius*), crested gecko (*Correlophus ciliatus*), and blue-tongue skink (*Tiliqua scincoides intermedia*).

NATURAL BEHAVIOR OF REPTILES

BEHAVIORS OF ALL REPTILES

- **Cold-Blooded:** All reptiles are *cold-blooded* (**poikilotherms**, ectotherms), i.e., they do not have a fixed body temperature, but they do have a preferred body temperature for maximum activity. They normally have a relatively low metabolic rate and eat infrequently compared to mammals.
  - **Preferred Body Temperature Purpose:** The preferred body temperature range enables the most activity, optimum reproductive functions, and the ability to digest food. The immune system is also most responsive at the preferred body temperature.

- **Essential for Health:** It is important to provide the correct environment that is appropriate for the species, which can vary from rain forest to desert.
- **Defense Tactics**
  - **Hiding:** Lizards and snakes can move quickly but only for short periods of time. They have little capacity for prolonged exertions. If unable to hide, snakes and lizards will attempt to rush to a nearby safer location.
  - **Coloration:** Coloration of their body aids in hiding in their natural habitat. Chameleons and anoles are able to vary their coloration to become less conspicuous.
  - **Chelonians' Protective Shell:** Tortoises and some turtles can retract their head and legs into their shells for defense.
  - **Defensive Roll:** If capture is eminent, many snakes and lizards will use a rolling maneuver to evade capture.
  - **Distraction Techniques:**
    - **Tail Sacrifice:** Many lizards and a few snakes are also able to shed the end of their tail, and the lost portion will continue to wriggle to distract an attacker.
    - **Faking Death:** Some snakes (eastern hognose snakes and grass snakes) will fake death to avoid attackers.
    - **Tail Waving:** Tail waving can be a distraction for predators from a snake's head and body to allow a biting strike by the snake to be more successful.
    - **Regurgitation:** Large snakes may regurgitate to distract a potential predator while the snake tries to escape.
    - **Urinating and Musking:** Lizards and turtles often urinate or secrete musk odors when picked up to distract a perceived predator, the handler.
- **Special Senses**
  - **Smell:** Hearing and sight are not a reptile's primary means of determining threats. Reptiles have a low respiratory rate, which does not move air quickly to smell odors in the way that mammals smell. Instead, snakes and lizards flick their tongue to be able to deliver odors to the vomeronasal organ rapidly for the perception of smell.
  - **Hearing:** Lizards have eardrums, but snakes and chelonians do not. Snakes and chelonians hear by feeling vibrations from the ground or in water and transmitting the information to their inner ears.
  - **Sight:** Lizards have upper and lower eyelids. Snakes have a transparent spectacle (eye covering), which clouds their vision during shedding.
- **Aggression**
  - **Intermale:** Adult male reptiles usually do not tolerate each other, particularly during breeding seasons. They associate rapid movement with aggression.
  - **Territorial Intrusion:** Chelonians and lizards are more territorial than snakes. Resentment of invasion of personal space or territory can cause eliminations, such as urine or foul-smelling musk.
  - **Body Language Depicting Agitation:** Aggression and defense are demonstrated in lizards, snakes, and some chelonians by elevating the body, open-mouth threats, vocalizations, tail flicking, and head bobbing.

CHELONIAN BEHAVIORS

- **Hiding:** Chelonians protect themselves by hiding when possible.
- **Shells:** A unique form of defense for chelonians is to draw a portion or all of their feet, head, and tail within their shell.
- **Camouflage:** Coloring to match environment colors provides some additional protection. Bottom dwellers, like alligator snapping turtles, have spikes on their *carapace* (upper shell) to catch algae for camouflage.

SNAKE BEHAVIORS

- **Defenses:**
  - **Hiding:** Snakes will usually attempt to avoid perceived danger by hiding.
  - **Alternate Defenses:** If this is not an option or is ineffective, some species will fake death, secrete foul odors, make threatening noises, flick or shake their tail, or inflate their body with air or spread the skin on their head to appear larger. For example, hognose snakes, when frightened, can inflate their bodies, flatten their necks, raise their head, and hiss, appearing like a cobra.
  - **Inherent Aggressiveness:** Some snakes are usually gentle, such as ball pythons, corn snakes, western hognose snakes, and gopher snakes. Kingsnakes may or may not be gentle. Other snakes may be typically bad-tempered, such as water and bull snakes.
- **Persistent Alertness and Special Senses:**
  - **Smelling:** A snake's tongue is located in a sheath in the front of the mouth. It is flicked to obtain chemical particles and deposit them in the vomeronasal organ in the roof of the mouth to determine the smell of the particles. Flicking the tongue indicates alertness.
  - **Vision:** Vision is best in arboreal snakes. Better vision is required for them to hunt birds. Burrowing snakes have poor vision.
  - **Hearing:** Snakes do not have ear canals or eardrums. Low-frequency sound vibrations are transmitted via their body, particularly their jaw when it is on the ground, to the columella in their inner ear.
- **Ingestion of Food:**
  - **Food:** All snakes are carnivorous. Most eat rodents.
  - **Swallowing:**
    - **Disarticulating the Jaw:** They swallow their food without chewing. Food that is 3 times the diameter of a snake's head can be swallowed because they can disarticulate their jaws. The limiting factor in size is how much the mouth and throat can stretch.
    - **Breathing While Swallowing:** Digestion of whole food is a slow process (2 or more days) in their GI tract. The trachea opens just behind the tongue sheath in the front of the mouth. This allows the snake to breathe when it has its mouth full of prey during the slow swallowing process.

LIZARD BEHAVIORS

- **Defenses:**
  - **Hiding:** Lizards prefer to hide for defense.
  - **Alternative Defenses:**
    - **Aggressive Posturing:** Lizards that cannot hide may resort to aggression. Bobbing of the head up and down in lizards is a warning of possible aggressive defense. When threatened, a bearded dragon will flare out the skin of its throat, gape its mouth, and bob its head up and down to bluff its threat to move away.
    - **Autonomous Tails:** Some lizards have autonomous tails that can be released while making an escape. A regrown autonomous tail is often different from the lizard's body color.
    - **Faking Injury:** Horned lizards can constrict the muscles in their neck to elevate their blood pressure enough to rupture small blood vessels near their eyes. The result is squirting blood from their eyes toward their perceived threat.
    - **Camouflage:** Chameleons move very slowly. Rather than hurrying to hide, they change their body color in an effort to become inconspicuous.

- **Shedding of Skin:** Lizards shed skin in patches.
    - **Normal Shedding:** Normal shedding (*ecdysis*) takes 5–7 days.
    - **Abnormal Shedding:** Failure to shed skin completely (*dysecdysis*) can allow the drying dead skin to constrict around the toes and end of the tail. Dysecdysis can be caused by lack of sufficient humidity or poor nutrition.

SAFETY FIRST

HANDLER SAFETY

No reptile is domesticated. Some tolerate being handled, but all prefer not to be handled and persist in seeking escape.

- **Potential for Injury:**
    - **Chelonians:** Chelonians can inflict tissue-crushing bites with the horny plates of their beaks, but with the exception of snapping turtles, most do not.
    - **Squamates:** Snakes and carnivorous lizards have teeth, and their bites produce penetrating wounds. Their upper teeth slant back toward their throat and will either tear tissue if the victim jerks away, or the reptile's teeth will dislodge in the wound. Nonvenomous pet snakes have two rows of upper teeth on each side and one row of lower teeth on each side.
- **Recognition of Common Dangerous Species:** Recognition of dangerous reptiles is important for avoidance or special handling.
    - **Aggressive Squamate Species:** Rock pythons, anacondas, Tokay geckos, and iguanas tend to be aggressive.
    - **Venomous U.S. Species of Squamates:** Gila monsters, Mexican beaded lizards, rattlesnakes, copperheads, cottonmouths (water moccasins), and coral snakes are venomous lizards and snakes in the U.S.
    - **Size:** All constrictor snakes and lizards longer than 5 feet are dangerous. If larger Squamates are handled, more than one handler should be involved.
- **Safer Handling of Chelonia**
    - **Low-Risk Chelonians:** Chelonia do not have teeth, but they have a beak with serrations that can inflict injury, and many have sharp claws. Tortoises generally hide in their shells and are not aggressive.
    - **Higher-Risk Chelonians:** Snapping, softshell, and mud turtles have long necks, dangerous bites, and bad dispositions. If restraint is absolutely necessary, a handler should hold these species by grasping the shell between the hind legs or grasp the top shell edge above the head and hold the base of the tail. They should not be picked up by their tails.
- **Safer Handling of Squamata**
    - **Inherently Dangerous Squamates:** Some snakes and lizards can grow to longer than 6 feet and become a physical danger to handlers. Constricting snakes longer than 8 feet and venomous snakes are inherently dangerous.
    - **Risk Control:**
        - **Train Young for Handling:** Snakes and lizards should be briefly handled daily when young. Handling must be performed slowly and gently. Handling is best done in early morning or late afternoon when the Squamate is possibly drowsy.
        - **Control of Environment:** Snakes and lizards should be moved to a separate, quiet, warm enclosure with no substrate to eat. Food should never be presented by hand.

- **Approach to Capture:** When taking the lid off any enclosure of an aggressive lizard or snake, the lid should be opened on the off side (away from the handler) and the lid held in front of the handler as a shield. Opening the side next to the handler may invite a strike or escape attempt toward the handler.
- **Low-Confrontation Handling:** When holding calm lizards or snakes, they should be given a chance to hold onto the handler's hands and arms more than the handler holds onto the lizard or snake. The goal should be to allow lizards and snakes being held to have the illusion of being free while protecting them from falling and preventing their escape. Squeezing their body and attempting to prevent them from moving may cause the lizard or snake to try to escape and bite the handler.

- **Potential Hazards of Handling Snakes**
  - **Temperament of Popular Pet Snakes:** Corn snakes and ball pythons are docile and popular pets.
    - **Ground Snakes:** Ball pythons are nocturnal ground snakes with a calm disposition.
    - **Arboreal Snakes:** Arboreal snakes, such as corn snakes, are more rapid moving than ground snakes. However, it is easier to accidently drop a ground snake, since they do not hold on to a hand or arm as well as arboreal snakes.
  - **Irritability during Shedding:** About 2 weeks prior to shedding, a snake produces a cloudy liquid between layers of the skin. This is most evident in the cloudy appearance to the eyes. During this period, snakes become irritable and likely to bite.
  - **Snakebites:** Snakes that cannot hide protect themselves by biting.
    - **Indication of Impending Bite:** Slowing of the flicking rate of the tongue and a stiffened body signals an impending strike.
    - **Food Triggers** (Table 7.2):

Table 7.2 Food Triggers for Snakebites
• Handlers should always wash their hands before handling snakes, especially if they have recently handled a snake's prey, which in some cases may be another snake.
• Feeding a snake in its primary enclosure may cause the snake to become aggressive whenever approached, so food should only be offered in a separate enclosure dedicated for feeding.
• Handling of snakes should be done at random times. If all handling is only at the time of feeding, snakes will associate hands with food and may become aggressive when handled.

- **Other Odor Triggers** (Table 7.3):

Table 7.3 Nonfood Odors That Trigger Snakebites
• The odors of fingernail polish or hand lotion may elicit a defensive response from a snake unadapted to the smell.
• A handler should never allow any snake near his face, since the odor of food can encourage some snakes to strike.

- **Consequences of Nonvenomous Snakebites** (Table 7.4):

Table 7.4 Consequences of Nonvenomous Snakebites
• Snake mouths contain many harmful bacteria, and all snakes are carnivorous with sharp teeth.
• Snake teeth angle backward, toward their throat, to aid in preventing the escape of their prey, and forceful withdrawal from a bite can rip victim's tissue, exacerbating the injury.

- **Removal of a Biting Snake:** If a handler is bitten on the finger or hand and the snake does not let go, the snake's head must be pushed forward while prying the jaws open to reduce tearing bitten skin further. Alternatively, a bitten handler may put a wooden spatula or plastic credit card in the snake's mouth and push between the handler's skin and the snake's teeth. Submerging a snake's head in water to stimulate it to release is unreliable, since they have air reserves in their air sacs and can hold their breath for a long time.
- **Delivery of Venom** (Table 7.5):

Table 7.5	Factors Affecting the Delivery of Snake Venom
•	Only four species of snakes are venomous in North America: rattlesnakes, copperheads, water moccasins, and coral snakes, but each has a unique venom.
•	Venomous snakes in captivity are more dangerous than those in the wild because captive snakes will store more venom than those that must hunt and kill their prey.
•	Coral snakes have the most potent snake venom in the U.S., but their venom delivery is poor in comparison to pit vipers, requiring a more prolonged bite and chewing action.
•	Venomous snakes are capable of metering the volume of venom delivered based on several factors, such as whether the bite is predatory or defensive, the duration of the bite, and the size of the target.

- **Constriction:** Constricting snakes constrict to kill food and may constrict on a perceived predator, including a handler.
  - **Danger of Bite:** A constrictor snake should never be put around a handler's neck. The chance of surviving the bite of most venomous snakes is greater than surviving a large constrictor snake attack. Large constrictor snakes can kill within a few minutes without the human victim being able to get to call out for help.
  - **Danger of Asphyxiation** (Table 7.6):

Table 7.6	Danger of Asphyxiation by a Constrictor Snake
•	If a snake constricts around the handler's neck, it is very possible that the handler will be unable to remove it by himself, especially if the snake is more than 5 to 8 ft in length.
•	Boids will bite the victim, anchor the tail on a stationary object, which prevents the victim from unwrapping the snake, and constrict on the victim.
•	Struggling and an increased respiratory rate will intensify the snake's strength of constriction. If help is available, the snake must be unwound from the tail, since pulling on the snake's head will cause more struggling and increase the strength of constriction.
•	When larger snakes are handled, an assistant handler should be present for every 5 feet of the snake's length.

- **Potential Hazards of Handling Lizards**
  - **Defensive Tactics:**
    - **Biting, Clawing, and Slashing:** Lizards defend themselves by biting, clawing, or slashing with their tail when they cannot escape what they perceive as a threat. Some have sharp dental plates for plant diets, and others have sharp teeth for eating insects or animals. All lizards may bite, some have long talons, and some have muscular tails that can inflict serious injury.
    - **Aggressive Lizards:** Iguanas have long, sharp claws and long tails that they lash with for defense. Monitors will bite and hold while thrashing with their bodies.
    - **Venoms:** The Gila monster and Mexican beaded lizard are the only venomous lizards in North America.
  - **Temperament of Common Pet Lizards:** Popular pet lizards are bearded dragons and leopard geckos. Leopard geckos are nocturnal, gentle, and quiet,

especially during the day. Bearded dragons, particularly males, can be aggressive to other bearded dragons, but they are docile to humans and easy to handle.

- **Food Aggression:** Lizards, like snakes, should not be fed in their primary enclosure to reduce the problem of aggression in anticipation of food each time a hand enters the main enclosure.

REPTILE SAFETY

- **International Illegal Wildlife Trade:**
  - **Risk–Benefit Ratio:** The international illegal wildlife trade is more lucrative than drug smuggling. It is easier to bribe wildlife and customs officials in many countries than drug enforcement officials and to alter documents, such as faking captive breeding. The risks of being caught are low, and existing fines are insufficient deterrents.
  - **Preferred Wildlife:** The preferred animals for smuggling are reptiles, since most are small, resilient, and require infrequent access to food and water. Rare species command very high prices.
- **Escape and Abandonment:** Pet reptiles have an extraordinary ability to escape, and many that do not will be abandoned by their owners. Several states, particularly in the southern U.S., are trying to deal with the problem of exotic reptiles escaping or being abandoned and released into the environment.
  - **Biopollution in South Florida:** In south Florida, 26% of all fish, reptiles, birds, and mammals are exotic. Feral Burmese and reticulated pythons are particularly a problem. They are ambush predators that have a bad disposition and survive well in environments like the cypress swamps of southern Florida.
  - **Importation or Interstate Transport of Pythons and Anacondas:** In 2012, the U.S. Fish and Wildlife Service listed the Burmese python, yellow anaconda, and Northern and Southern African pythons as injurious invasive species under the Lacey Act. This makes it a federal crime to import these snakes or transport them across state lines.
- **Chelonian Safety—Turtles, Terrapins, Tortoises:** Aquatic chelonians have smaller *plastrons* (lower portion of the shell) compared to terrestrial species and are often unable to completely withdraw their head, neck, and limbs into the shell. To compensate for this, aquatic chelonians, such as softshell turtles and snapping turtles, are aggressive and capable of inflicting severe bites.
- **Snake Safety**
  - **Feeding Safety**
    - **Avoid Handling after Feeding:** Snakes should not be handled within 24 to 48 hours after feeding. Otherwise, they may regurgitate as a defensive tactic and become malnourished.
    - **Avoid Live Food:** Live food (mice/rats) should not be released in a box with a snake because the snake may be injured, especially if it is in the process of shedding its skin. Feeding of live prey to snakes is illegal in some European countries. Rodents as food should be humanely prekilled and from captive colonies that are disease and parasite free.
    - **Feed Individually:** Snakes should also not be fed in groups, since competition for the food may cause injuries.
    - **Use Tongs to Handle Food:** Tongs should be used to provide the food to keep human scent separated from the snake's thoughts of food.
  - **Risks during Shedding**
    - **Normal Shedding:** Young snakes shed their skin about once per month. The frequency decreases with growth, and adult snakes shed about twice per year. Shedding in snakes begins by stretching the mouth open, and typically the entire body from front to back will shed in one piece.
    - **Avoid Handling during Shedding:** Snakes should not be handled when their eyes are clouded by shedding skin. They do not eat, cannot see well,

and will become agitated during shedding. The new skin of snakes just following shedding is fragile and can be damaged.

● **Musculoskeletal Injuries to Snakes during Handling**
  – **Tail Muscle Injury:** Handlers should never restrain a snake by its tail, since there is risk of muscular injury to the snake.
  – **Bone Fractures:** Ball pythons and corn snakes are bred strictly for unusual colors by some breeders, using inbreeding and subsequently weakening the species. Snake diets should also be assessed, since brittle bones from malnutrition are common and require gentle handling to avoid fractures.

● **Lizard Safety:** Before handling lizards, handlers should be familiar with their species, temperament, and diet.

  ● **Metabolic Bone Disease:** In pet lizards, poor diets can cause a metabolic bone disease that is characterized by demineralized bones that fracture easily. In addition, most lizards have explosively quick movements that can put them in danger of being dropped during handling. Proper handling restraint and handling over tables can minimize the risk of the lizard being dropped and possibly breaking bones from the fall.

  ● **Injuries from Live Prey:** Live insect prey should not be left in the feeding enclosure. If the lizard loses interest, the insects may cause eye damage to the lizard by feeding on eye moisture.

KEY ZOONOSES

(**Note:** Apparently ill animals should be handled by veterinary professionals or under their supervision. Precautionary measures against zoonoses from sick animals are more involved than those required when handling apparently healthy animals and vary widely. The discussion here is directed primarily at handling apparently healthy animals.)

The risks of zoonoses from wild-caught reptiles are much greater than from captive-bred reptiles (Table 7.7).

Table 7.7 Diseases Transmitted from Healthy-Appearing Captive-Bred Reptiles to Healthy Adult Humans					
Disease	Agent	Means of Transmission	Signs and Symptoms in Humans	Frequency in Animals	Risk Group*
Bites and clawing	——	Direct injury	Bite wounds to face, arms, and legs	All reptiles are capable of inflicting bite or claw wounds or both.	3
Salmonellosis	*Salmonella arizona, S. marina,* among others	Direct, handling reptiles and indirect from fomites (cages, bowls, cage toys)	Diarrhea, systemic disease, and abscesses	Common	3
Edwardsiellosis	*Edwardsiella tarda*	Direct, handling reptiles and indirect from fomites (cages, bowls, cage toys)	Diarrhea and wound infections with gangrene	Common	3

* Risk Groups (National Institutes of Health and World Health Organization criteria. Centers for Disease Control and Prevention, Biosafety in Microbiological and Biomedical Laboratories, 5th edition, 2009.)
1: Agent not associated with disease in healthy adult humans.
2: Agent rarely causes serious disease, and preventions or therapy possible.
3: Agent can cause serious or lethal disease, and preventions or therapy possible.
4: Agent can cause serious or lethal disease, and preventions or therapy are not usually available.

SANITARY PRACTICES

Other than bites and claw wounds, the only zoonotic disease of great significance from captive-bred reptiles is salmonellosis. There is a high degree of risk of acquiring salmonellosis from reptiles, including those that appear healthy. The morbidity and mortality of salmonellosis can be high, particularly in humans who are young, elderly, or otherwise have impairment of their immunity.

- **Reduce Risk from Enclosures**
  - **Food Preparation Areas:** Reptile enclosures should not be located in or near human food preparation or storage areas. Reptiles should not be allowed to roam freely in a home or living area, and they should be kept out of food preparation areas.
  - **Routine Cleaning and Disinfection:**
    - **Daily Spot Cleaning:** Enclosures should be spot cleaned daily, with a thorough cleaning on a regularly scheduled basis.
    - **Disinfection of Enclosures:** Cleaning of enclosures should include disinfection with 5% bleach (sodium hypochlorite) followed by thorough rinsing before reintroducing the animal. Phenol or pine-scent disinfectants should be avoided.
    - **Clean and Disinfect Cleaning Equipment:** All cleaning equipment such as sponges, buckets, and sinks should be cleaned and disinfected. Cleaning reptile enclosures should not involve soaking in bathtubs, basins, or laundry sinks. When cleaning reptile enclosures, gloves and protective glasses or goggles should be worn.
- **Handlers' Personal Protection:**
  - **Appropriate Attire:** A handler of reptiles should wear appropriate dress to protect against skin contamination with skin scales or saliva, urine, and other body secretions.
  - **Personal Sanitary Practices:** Reptiles should never be fed by hand or allowed near a human's face. Hands should be washed after handling any reptile or objects touched by the reptile. Handlers should not eat or drink while handling reptiles.
  - **Avoid Handling if Immunoincompetent:** Young children, the elderly, pregnant women, and people with immunosuppressive diseases or on immunosuppressive medications should not handle reptiles due to risk of salmonellosis.

TURTLES AND TORTOISES (CHELONIANS)

APPROACHING AND CATCHING

- **Size Considerations:**
  - **Small Chelonians:** Small chelonians can be captured by grasping the sides (bridge) of their shell from above with the handler's thumb between the front and hind legs on one side and the first three fingers on the other side between the other side's front and hind legs (Figure 7.1). Putting a wrap of self-adhering elastic bandage material around the shell can provide a better gripping surface, if needed.
  - **Large Chelonians:** Large chelonians can be captured with two hands by the handler placing his thumbs on each side of the carapace (the upper shell) and other fingers under the plastron (the lower shell). Towels should be used to handle aquatic species for protection against their long claws.

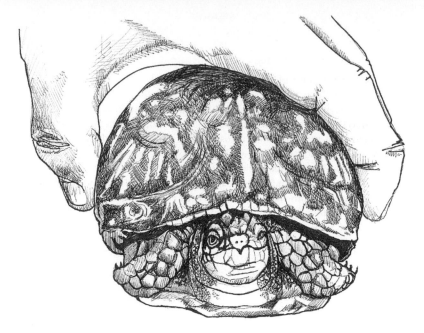

Figure 7.1   One-hand restraint of a small turtle.

- **Aggressive Chelonians:**
  - **Species:** Common snapping turtles (*Chelydra serpetina*), alligator snapping turtles (*Macroclemys temmincki*), and soft-shelled turtles can extend their head a long distance (2/3 the length of their shell) to inflict a bite.
  - **Approach:**
    - **Gain Control of the Tail:** Aggressive chelonians can be grasped by their tail for initial control.
    - **Lifting Holds:** To lift the turtle, its tail is firmly held and its body supported by grasping the edge of the carapace (upper shell) behind the neck or a hand placed under the back part of the plastron. Small biting turtles can be grasped and lifted by holding the back portion of the shell along with both hind legs.
  - **Protection from Claws:** Soft-shelled turtles will scratch handlers with their claws and are difficult to hold without injuring them. Handlers should wear light gloves to protect themselves from scratches.

HANDLING FOR ROUTINE CARE AND MANAGEMENT

- **Inversion Caution:** If a turtle needs to be turned over for examination, inversion must be done slowly to reduce risk of intestinal torsion. Other than for brief examinations, handlers should not hold a turtle or tortoise upside down or with the chelonian's head lower than its heart. Either of these positions makes it difficult for the chelonian to breathe properly.
- **Physical Examination:** If examination of the chelonian's head is needed, the handler should gently pull on a foreleg or gently prod near the rectum to cause the head to come out and then quickly grasp it with the handler's fingers behind the jaws. The head may also be able to be lured out by offering fresh fruit, especially berries. Some tortoises and some turtles can retract their head and feet so effectively in their shells that chemical restraint is needed to handle these body parts for examination or treatment.

APPROACHING AND CATCHING

- **General Considerations of Approach:**
  - **Captive Born versus Wild Caught:** If pet snakes have been raised in captivity and handled gently while young, they are usually easily handled with little restraint. Garter snakes, kingsnakes, hognose snakes, and gopher snakes caught in the wild are sometimes kept as pets, but these are more intolerant of handling, susceptible to stress, and may have diseases and parasites they acquired in the wild.
  - **Snakes More Than 5 Feet in Length:** Snakes kept as pets should not exceed 5 feet in length, as they require more than one handler to be present when handled. Constrictors more than 8 feet in length are so strong that they are considered inherently dangerous. Many escape or are released when they become a burden or bore to their owner, resulting in endangerment to the snake's health and survival, indigenous wildlife, or unsuspecting humans.
- **Catching a Tame Snake (*Procedural Steps 7.1*):**

Procedural Steps 7.1	Catching a Tame Snake
1.	A handler should make sure the snake is aware of his presence and move at slightly slower-than-normal speed, because snakes bite in self-defense or a feeding frenzy.
2.	A tame snake is picked up by placing a hand toward the snake's side with outstretched fingers and slid under the first 1/3 of the snake's body.
3.	As it is picked up, the remainder of the body should be supported with the handler's other hand with fingers spread to provide wider support (Figure 7.2).
4.	The snake's head should not be reached for first nor the body held so tight that it cannot keep moving.

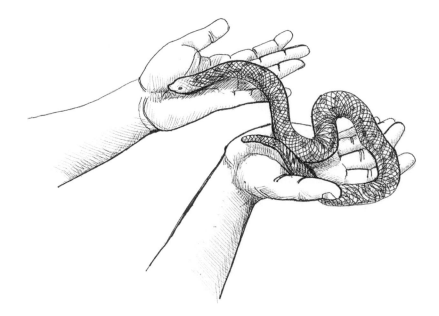

Figure 7.2 Holding tame snakes should not inhibit their movement.

- **Aggressive Snakes:** Young snakes, shedding snakes, and snakes expecting food tend to be more likely to bite. Arboreal snakes will try to progress up the handler's arm which, if permitted, can allow proximity to the handler's face and neck. This should not be allowed.

HANDLING FOR ROUTINE CARE AND MANAGEMENT

Snakes are typically supported with their movement directed, not held in a manner to inhibit their movement.

- **Allow Movement While Being Held:**
  - **Muscle Injury from Excessive Restraint:** Holding them tightly stimulates the snake to attempt escape from a predator. This can also seriously damage their muscles and cause death days later. They should be given the illusion that they are free to escape when they want.
  - **Rolling Hands:** As they are loosely held, a *rolling hands* technique of holding them gives them the illusion that they are not trapped. They are allowed to move from one hand to the next, then the hand they left becomes the next hand they move to. Immobilizing types of restraint should not typically be used.
- **Never Hold near Handler's Face:** When holding a snake, it should never be held near the handler's face. Bites could occur from defense, aggression, or food odors.
- **Avoid Territorial Aggression:**
  - **Reaching into an Enclosure:** The most likely time for a handler to be bitten is when reaching into the snake's enclosure. The handler may startle them or because of the movement of his hand being perceived as food or containing food, especially if the handler has snake food odor on his hands.
  - **Blocking Technique:** When reaching into an enclosure for an unfamiliar snake, the handler should block the snake's head with one hand held flat with fingers together. The purpose is to create a barrier over the snake's head while reaching for the body with the other hand. A flat hand is more difficult to bite.
- **Avoid Food Aggression and Regurgitation:** Handling should not be attempted if food smell is on the handler's hands or if the snake has recently eaten. Food odor may stimulate the snake to strike. Handling a snake soon after it has eaten may cause the snake to regurgitate.
- **Techniques for Potentially Dangerous Snakes:**
  - **Snake Hook:** If a snake is possibly dangerous, a snake hook first should be used to lift the snake and then grasp its body.
  - **Pinning the Head:** For those snakes that are known to be dangerous, the head should be immobilized before picking the snake up.
    - **Grasping the Head:** The basic hold for snake head restraint is to grasp the base of the skull between the thumb and middle finger with the index finger on top of the head.
    - **Pinning Hook:** A snake pinning hook may be needed to pin the neck down on a soft surface to limit movement until the snake's head can be grasped for manual control.
  - **Restraining the Body:** The snake's body should be restrained and supported after capturing the head to prevent thrashing and breaking its back. The snake's head should be held firmly without being squeezed (Figure 7.3). Approximately one handler is needed for each 5 ft in the length of a snake to control boids.
- **Anticipate Musking:** Snakes have musk glands near their cloaca that they may use to excrete a malodorous secretion, which is also distasteful to their predators.

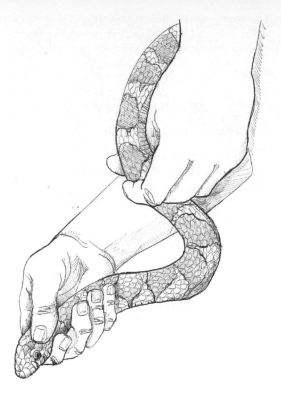

Figure 7.3 Dangerous snakes must be held by their head, and their body should be supported with the handler's other hand.

SPECIAL EQUIPMENT

- **Lifting Hook**
  - **Purposes:**
    - **Moving and Lifting:** Snake hooks can be used to move snakes a short distance such as into a transport bag. The hook is worked under the snake between the first one-third to one-half of the snake's length to pick it up. The snake will remain still, trying to keep its balance.
    - **Guiding Movements:** Hooks can also help in guiding the movement of snakes on the ground or a floor.
  - **Working Position—Always Tilt Down:** Snake hook poles should always be tilted down, away from the handler. Otherwise, the snake may slide toward the handler.
- **Pinning Hook**
  - **Purposes:** A pinning hook is a Y-shaped stick with tubing for padding can be used to introduce a handler's presence and, if necessary, pin a snake's head. Pinning sticks should be used when the snake is on a padded surface to reduce risk of injury to the snake.
  - **Method (*Procedural Steps 7.2*):**

Procedural Steps 7.2   Capture of Snakes with a Pinning Hook
1.   The snake's head is immobilized by pressing the elastic band just behind the head, pushing the head down and trapping it until the head can be grasped by a free hand. Two pinning sticks may be necessary for difficult snakes.

2.	The base of the head is then grasped between the thumb and middle finger with the index finger on the top of the head and the stick removed.

- **Shields and Squeeze Box:** Plexiglass or wire mesh shields with handles can be used to pin snakes until the head can be restrained. Properly fitted ventilated plexiglass or wire mesh lids on a box can be used to contain movement of the snake while the shield descends into the box to squeeze them for administration of injectable medication or chemical restraint. The bottom of the squeeze box should be padded.
- **Capture Poles**
  - **Improvised Capture Pole:** A capture pole can be made with a 3-foot-long wooden pole, eye screw, and a long cord. One end of the cord is tied or otherwise fixed to the end of the pole. The other end of the cord is run through an eye screw placed an inch from the end of the pole where the end of the cord is fixated, creating a capture loop between the fixed end and the eye screw.
  - **Technique:** The capture loop is dropped around the snake's neck and the loop closed on the neck by pulling on the cord. The risk of injuring the snake is greater than with a hook, but if appropriate pressure is applied with the loop and restraint is short, a capture pole can be safe and effective.
- **Capture Tongs:** Capture tongs are long-handled metal grasping instruments.
  - **Limitations:** It is difficult to gauge the pressure being exerted with tongs, so the risk of injury to a snake can be significant. Tongs can make a snake thrash and bite itself. Capture tongs should not be the sole means of restraint of a snake, but tongs can be useful to assist with handling with a hook.
  - **Grasping Food or Environmental Enrichments:** Tongs should be used in presenting food to large snakes and in moving environmental enrichment objects in a snake enclosure.
- **Transparent, Flexible Tubes**
  - **Purpose:** Bad-tempered snakes, such as small reticulated pythons, with a history of inappropriate biting should be handled in the same manner as venomous snakes using transparent, flexible tubes.
  - **Method** (*Procedural Steps 7.3*):

Procedural Steps 7.3	Tube Restraint of Aggressive Snakes
1.	Aggressive snakes are moved to a large plastic bucket with a transport bag or using a snake hook.
2.	A snake hook can help guide the head if needed. Use of a cone to guide the snake into the tube is another method.
3.	As the snake investigates a possible escape route upward and out of the bucket, a flexible, preferably darkened, tube can be placed over the snake's head and down part of the front of its body.
4.	The tube should be just large enough to accommodate the thickest part of the snake's body so that it cannot turn around in the tube and long enough to keep the handler's hand on the tube, out of danger.
5.	When the snake has entered one-third of the tube's distance, the snake and tube are grasped to entrap the snake (Figure 7.4).
6.	If the first tube seems too large in diameter, a smaller tube can be slid down the open end to the snake for it to enter, the snake and small tube are grasped, and then the larger tube is removed.
7.	Releasing the snake back into the bucket, transport bag, or enclosure is done by allowing it to move forward through the tube and out the other end.

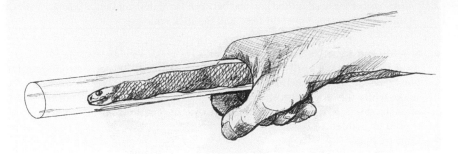

Figure 7.4   Tube restraint for dangerous snakes.

LIZARDS

APPROACHING AND CATCHING

- **General Considerations:**
  - **Easy to Handle:** Leopard geckos and bearded dragons are the easiest lizards for new owners of lizards to handle and manage. Other common small lizards that are relatively easy to handle include anoles, skinks, and chameleons.
  - **Hazardous to Handle:** Green iguanas (*Iguana iguana*) are common lizard pets even though they are territorial, aggressive, have long claws, will bite, and have a long, muscular tail which they use to lash handlers if excited. Only experienced handlers should handle large iguanas.
- **Small to Medium-Sized Lizards:**
  - **Method (Procedural Steps 7.4):**

Procedural Steps 7.4 Catching a Small to Medium-Sized Lizard	
1.	Small to medium-sized lizards (geckos, bearded dragons, uromastyx) accustomed to being handled can be grasped from above their body.
2.	The handler should move his hand slightly slower than normal to capture lizards.
3.	A lizard's shoulders and the pelvic area should be supported and restrained as needed to prevent the lizard from lashing and damaging its vertebrae (Figure 7.5).

  - **Control Lighting:** Subdued lighting is helpful. Many lizards will become more difficult to handle and aggressive if in natural sunlight.
- **Iguanas:**
  - **Control the Tail:** Well-handled iguanas may tolerate moderate handling without resistance. Handlers must exercise special care to control an iguana's long, muscular tail. A defensive position they may take is to bend in a U shape to ready the tail to slash in defense.
  - **Risk of Snaring the Head:** In some cases, snares may be used to gain control of the head. When this is done, the tail must be quickly grasped as soon as the head is snared to prevent thrashing that could break the iguana's neck.

Figure 7.5 Small lizards can be held loosely.

- **Method for Calm Iguanas (*Procedural Steps 7.5*):**

Procedural Steps 7.5	Catching Calm Iguanas
1.	Approach the lizard from one of its sides.
2.	Reach over its body and place a hand under its chest.
3.	A natural tendency of arboreal lizards is to grab hold of the closest surface or object if being picked up.
4.	The feet should be promptly pulled loose with care not to injure the lizard's legs or feet while its body is lifted.
5.	Support the front of the body with the hand and forearm underneath its chest.
6.	The hand is held between the front legs with one finger between the legs.
7.	The abdomen and pelvis are supported on the forearm, the hind legs lie on both sides of the forearm, and the upper arm pins the tail next to the handler's body (Figure 7.6).
8.	Larger lizards require the other hand to support the hindquarters (a two-hand hold) rather than the forearm of the hand supporting the chest and shoulders (Figure 7.7).
9.	The head is held directed away from the handler to avoid being bitten.

- **Method for Agitated Iguanas (*Procedural Steps 7.6*):**

Procedural Steps 7.6	Catching Agitated Iguanas
1.	Signs of iguana agitation include sharply blowing air through the nose, pushing a hand away, going into a C posture to prepare for a tail slap, and if being picked up, doing a sudden *crocodile roll*.
2.	Capture of a defensive iguana involves grasping the base of the tail with one hand, lifting the hind legs off the floor, and then grasping the neck and shoulders with the other hand. Immediately after picking the iguana up, the handler should trap the tail between his forearm and body.
3.	Alternatively, the iguana can be captured using the *taco technique* with a towel and then wrapped in the towel like a burrito for restraint.

Figure 7.6 Support for carrying a small to medium-sized lizard.

Figure 7.7 Support for carrying a large-sized lizard.

HANDLING FOR ROUTINE CARE AND MANAGEMENT

- **Loose Restraint Is Generally Preferred:** Lizards are most comfortable if they can continue to move, or think they can move, at their own will. Holds that primarily provide support while the handler directs their continuing movements are most successful. Small, calm lizards are best held loosely and by letting the lizard move from one hand to the other in a rolling hand movement.
- **Common Handling Equipment:** A blanket or towel to cover a lizard's head may be used to assist in its capture and immobilization. Lizards used for research may be restrained with small clear plastic tubes for brief physical restraint similar to restraint tubes for rodents.
- **Small Lizards:** Unnecessary handling and restraint should be avoided whenever possible.
  - **Minimum Restraint:** Small, docile lizards can be easily picked up and held in a palm and on an arm without any restraint. Care must be taken since small lizards that are not handled often may attempt to jump off the handler's hand. Tiny lizards like anoles or small geckos can be easily injured even with careful handling.
  - **Risk of Tail Damage:** When additional restraint is needed, a small lizard should never be restrained by its tail, which will break off, a process called *tail autonomy.* Most lizards are very fast and should be grasped around the shoulders and pelvis. This can be done with one hand for smaller lizards.
- **Iguanas and Other Large Lizards:** Arboreal lizards, such as green iguanas, have long claws to help them climb. Their claws can inflict serious injuries to handlers if the restraint applied does not prevent them from being able to rake their claws on the handler.
  - **Avoidance of Injuries from Claws**
    - **Trimming Claws:** Anytime a large lizard is captured, consideration should be given to trimming the nails to reduce risk of handler injury with further restraint. A blinder wrap can be helpful in trimming nails. The wrap is created by padding the eyes with cotton balls and wrapping the head with self-adhering elastic bandage material.
    - **Padded Sleeves:** When handling small to medium iguanas, a jacket or coat with thick sleeves should be worn to protect forearms from the long claws.
    - **Gloves and Gauntlets:** Leather gloves with gauntlets and a jacket with long sleeves should be used if a larger lizard has long claws.
  - **Handling Techniques**
    - **Tame Iguanas** (*Procedural Steps 7.7*):

Procedural Steps 7.7	Handling Tame Iguanas
1.	If handled often, an iguana can be picked up by sliding a hand underneath its body and between its legs.
2.	A slow, quiet approach is best, for if the iguana becomes excited, it will grab whatever it can and hold on to prevent being picked up.
3.	The front end is supported with a hand below the thorax between the front legs, and the rest of the torso rests on the handler's forearm.
4.	The lizard's tail is restrained under the handler's arm and against the handler's body.

- **Resisting Iguanas** (*Procedural Steps 7.8*):

Procedural Steps 7.8	Handling Resistant Iguanas
1.	Another restraint for medium to large-sized lizards is to grasp them from above their back.

Procedural Steps 7.8	Handling Resistant Iguanas
2.	The thumb and index finger is placed on the lower neck and three fingers behind one shoulder.
3.	The other hand restraining the pelvis is positioned with an index finger in front of a hind leg, the thumb behind the other hind leg, and three fingers on the base of the tail.

- **Special Handling Equipment:** Towels can be used as hoods and wraps for capture when needed. A noose can be made of thick cord to snare a lizard that quickly evades hand capture. Nets can be used for capturing difficult cases.
- **Handler Injuries**
  - **Posturing Signals of Attack:** Posturing and other body language can signal aggression from a lizard. Defensive posturing can include tail whipping, head bobbing, opening the mouth wide, standing higher on all four legs, standing broadside, and extending their dewlap forward.
  - **Biting:** If a handler is bitten by a lizard, many lizards will not release, and their bite will intensify if the victim struggles. If being quiet and calm does not result in the lizard ending its bite, the lizard should be placed on the floor, with the victim lowering his body if necessary. The lizard's attention should shift to letting go and attempting to escape or assuming a defensive posture in an attempt to scare the victim away.
  - **Spraying Offensive Odors:** Other deterrent maneuvers by lizards can include spraying musk or urine and feces from the cloaca.

HANDLING OF REPTILES FOR COMMON MEDICAL PROCEDURES

GENERAL CONSIDERATIONS

- **Need for Chemical Restraint:** Typically, handling and restraint of reptiles can and should be done without tranquilization, sedation, hypnosis, or anesthesia. However, some handling and restraint procedures should be performed with adjunct chemical restraint or complete immobilization by chemical restraint if otherwise the safety of the handler or reptile would be in jeopardy.
- **Administration of Medication:**
  - **Routes:** Routes for injections are intramuscular, subcutaneous, intracoelomic, intravenous, or intraosseous.
  - **Injections Preferred:** Most medications to reptiles are given by injection rather than by mouth. Injections are preferred in larger reptiles due to danger to handlers when handling a reptile's head and mouth.
  - **Hydration and Optimum Temperature:** For injections other than intravenous to be effective, the animal must be well hydrated and at a preferred temperature for normal activity for its species.

INJECTIONS AND VENIPUNCTURE

Insertion of transcutaneous needles for injection or aspiration in reptiles carries the risk of slashing tissue beneath the skin, including damage to nerves and blood vessels, and breaking hypodermic needles off in its body. The area in which the needle is to be inserted must be immobilized, and the reptile's mouth and feet should be restrained from interfering with the procedure, especially venipunctures.

- **Venipuncture**
  - **IV in Chelonians:** Veins that can be accessed in chelonians include the jugular vein, brachial vein, dorsal and ventral coccygeal (tail) vein, femoral vein, and

subcarapacial (beneath the shell) vein, found on the midline just under the carapace (upper shell) above the retracted head of the turtle. The jugular vein on the right side is preferred for turtles, and the dorsal vein of the tail is preferred for tortoises. The head must be captured and extended to access the jugular vein.

- **IV in Snakes:** The vein used for venipuncture in snakes is the ventral coccygeal vein caudal to the vent on the midline of the tail. Chemical restraint may be necessary. In anesthetized larger snakes, the palatine vein in the roof of the mouth can be accessed.
- **IV in Lizards:**
  - **Eye Cover:** Restraint for venipuncture in lizards may be assisted by using cotton balls over its eyes and using self-adherent elastic bandage loosely wrapped around the head to hold the cotton in place.
  - **Sites:** The ventral coccygeal (tail) vein is often used and accessed with the lizard held ventrodorsal (on its back). However, use of the ventral coccygeal vein can be hazardous in restraining lizards with autonomous tails. The ventral (central) abdominal or jugular vein can be accessed after chemical restraint and ventrodorsal positioning, but hematomas often occur afterwards.
- **Intramuscular Injections:** Intramuscular (IM) injections are the most common method of drug administration to reptiles.
  - **Anterior Injection Sites:** It has been traditionally believed that IM injections given in the hind limbs or tail, are absorbed and carried to the renal portal system resulting in more rapid elimination from the body and uneven distribution in the body. Injections in the front half of the body of reptiles have been preferred. Although evidence is lacking for clinically significant first-pass elimination of drugs by the renal portal system in all cases, most veterinarians still make IM injections in the front half of the body of reptiles.
  - **IM in Chelonians:** In chelonians, IM injections are usually given in the upper (proximal) forelimb (Figure 7.8). Although less convenient, the pectoral muscles at the junction of neck and forelimb can also be used. The hind limbs may be used, but the medication dosage may need to be increased.

Figure 7.8 Intramuscular injection site in turtles.

Figure 7.9 Intramuscular injection site in snakes.

- **IM in Snakes:**
  - **Injection Sites:** Snakes are administered IM injections in the dorsal muscles of the back in the front one-third of the body (Figure 7.9). Injections should be performed by angling the needle toward the head and inserting the needle underneath scales, not through them.
  - **Hand Restraint:** No restraint or hand restraint of the head may be needed.
  - **Squeeze Box:** When full-body restraint is needed, a squeeze box can be used. A squeeze box for snakes is a box with a removable top (***Procedural Steps 7.9***).

Procedural Steps 7.9	Squeeze Box for Intramuscular Injections in Snakes
1.	To administer IM injections, the standard top is temporarily replaced with a slightly smaller wire mesh top with handles that can be slid down the box, pressing the snake on the bottom.
2.	An alternative to a wire mesh treatment top is 0.5 cm plexiglass drilled with multiple breathing and injection access holes and having a center handle.
3.	The bottom of the treatment box for squeezing snakes from above should have a firm but soft bottom, such as foam rubber.

- **IM in Lizards:** The deltoid muscles of the shoulders are preferred for IM injections in lizards (Figure 7.10). The forearm muscles of large lizards may also be used. Chameleons and small geckos do not have sufficient muscle for IM injections.
- **Subcutaneous Injections:** Subcutaneous (SC) injections are not often administered to reptiles.
  - **SC in Chelonians:** When they are, the injection in chelonians is given in the ventral neck flap or under the skin cranial to the fore- or hind limbs.

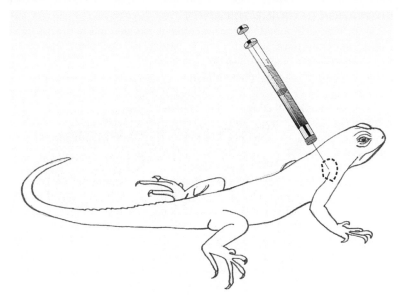

Figure 7.10 Intramuscular injection site in lizards.

- **SC in Snakes:** In snakes, the injection is given in the cranial one-third of the body in an epaxial (dorsolateral) area. As with IM injections, the needle should be angled toward their head and the injection made underneath, not through, the scales.
- **SC in Lizards:** Lizards are given SC injections in an epaxial thoracic area.
- **Intracoelomic Injections:** Intracoelomic (IC) injections are uncommonly used in reptiles. Reptiles do not have a diaphragm, so their body cavity is unlike the chest and abdominal cavities of mammals. IC injections can cause compression of their lungs, and there is risk of puncturing or cutting an internal organ with the injection needle.
    - **IC in Chelonians:** When IC injections are given to chelonians, they are administered through the caudal skin folds where they attach to the bridge (junction on their sides between the upper and lower shell).
    - **IC in Snakes:** In snakes, IC injections are given in the lateral part of the body just dorsal of the ventral scales and in front of the cloaca in the caudal one-fourth of the body.
    - **IC in Lizards:** Lizards are placed in ventrodorsal recumbency (on their back) with their head tilted downward. The injection is given in the caudal one-third of the body. The midline should be avoided due to the risk of penetrating the ventral abdominal vein.
- **Intraosseous Injections:** Intraosseous (IO, inside a bone) injection should only be done in reptiles after they are radiographed to check bone density. If they have metabolic bone disease, an IO puncture is likely to fracture the bone.
    - **IO in Chelonians:** Injections in chelonians are given either at the junction between the plastron and carapace just cranial to a hind leg or into the tibial crest.
    - **IO in Snakes:** Intraosseous injections are not possible in snakes.
    - **IO in Lizards:** In lizards, IO injections are performed on the proximal femur (entered between hip joint and greater trochanter) or the distal femur (entered at stifle joint). The proximal tibia may be used in larger lizards. The point of injection is at the tibial crest.

ADMINISTRATION OF ORAL MEDICATION

Giving drugs to reptiles by mouth can be hazardous to handlers due to the risk of being bitten or difficult due to the resistance of the reptile, particularly chelonians. Oral medication is used more often in lizards than in snakes or chelonians.

- **Adding to Food:** If possible, the best method of oral medication to reptiles is to apply the medication on or into food. The amount of food should be limited to ensure the likelihood the reptile will consume the medicated food.
- **Oral Syringe:** In some cases, an oral syringe with a metal ball-tipped gavage needle or rubber feeding tube can be placed in the corner of the mouth and the medication delivered slowly.
- **Feeding Tubes:**
    - **Caution Needed to Open Mouth:** Oral administration of medications with a lubricated feeding tube can be safe for the reptile if care is used to open a reptile's mouth with a wooden spatula or popsicle stick. In some cases, the lower neck flap (dewlap) can be pulled gently to open the mouth.
    - **Avoid Insertion into the Glottis:** It is relatively easy to avoid the glottis (opening to the windpipe). The glottis in snakes is located in the front of the mouth. This allows the snake to have its mouth stuffed with a meal and still breathe. It is at the base of the tongue in lizards and chelonians.
    - **Insert Tube to Distal Aspect of the Esophagus:** If a feeding tube is inserted too far, the stomach may be reached which can be hazardous to the reptile. The length from mouth to stomach should be estimated beforehand.
        - **Chelonians:** In chelonians, the appropriate distance is from the nose to the junction of pectoral and abdominal *scutes*. Scutes are plates in the shell of chelonians.
        - **Snakes:** In snakes, the distance is one-third of its body length.
        - **Lizards:** In lizards, it is one-half the body length.

TRANSPORTING REPTILES

CHELONIANS AND LIZARDS

Opaque plastic storage boxes or tubs are the best means of transporting chelonians and lizards. If a chelonian is an aquatic species, it should have a moist towel under and over it. Transporting lizards should be in plastic containers with tight lids and adequate ventilation.

SNAKES

- **Double-Lined Canvas or Cotton Bag**: When moving snakes, a double-lined bag should be used to contain the snake and the bag placed in a plastic box. The bag should be canvas or cotton to allow ventilation. Pillowcases (breathable cotton) will suffice for most nonvenomous snakes.
    - **Double-Stitched:** Seams should be double-stitched to prevent the snake's head from pushing a hole through a seam. Having a double-bottom or stitched-cornered bag will protect the handler's hand from a bite if the bottom of the bag is grasped while removing the snake.
    - **Tie Off After Inserting Snake:** Gentle snakes can be simply lifted into a transport bag. Transport bag openings should be tied off and the bag lifted and carried by the end of the knot to reduce the risk of being bitten through the bag.
- **Inserting Dangerous Snakes:**
    - **Triangular Metal Hooks:** For added safety in bagging aggressive or venomous snakes, the bag's neck can be placed through a triangular metal hoop at the

end of a pole. This will hold the bag open without putting the handler's hands at risk.
- **Snake Hook:** A snake hook is used to place the snake in the bag. The bag is removed from the hoop and the bag neck tied tightly with a cord.
- **Removing Snakes from Bags (*Procedural Steps 7.10*):**

Procedural Steps 7.10	Removing a Snake from a Transport Bag
1.	The handler locates and grasps the snake's head, using the bag as a shield.
2.	The handler then reaches into the bag and grasps the base of the head and then uses his other hand to lift the body as the snake is removed from the bag.
3.	Aggressive snakes are released from a head hold by removing the thumb and middle finger while the head is pressed down with the index finger, which is then immediately removed to complete the release.

NOTE

Additional recommended readings on reptile handling are available in references on multiple species of small animals provided in the Appendix.

REPTILE HANDLING REFERENCES AND RECOMMENDED READINGS

1. Ballard B, Cheek R. Exotic Animal Medicine for the Veterinary Technician, 3rd ed. Wiley-Blackwell, Ames, IA, 2017.
2. Bays TB, Lightfoot T, Mayer J. Exotic Pet Behavior. Saunders, St. Louis, MO, 2006.
3. Clancy MM, Davis M, Valitutto MT, et al. Salmonella infection and carriage in reptiles in a zoological collection. J Am Vet Med Assoc 2016;248:1050–1059.
4. Holz P, Barker IK, Crawshaw GJ, et al. The effect of the renal portal system on pharmacokinetic parameters in the red-eared slider (*Trachemys scripta elegans*). J Zoo Wild Med 1997;28:386–393.
5. Judah V, Nuttall K. Exotic Animal Care and Management, 2nd ed. Delmar, Cengage Learning, Albany, NY, 2016.
6. Woodward DL, Khakhria R, Johnson WM. Human salmonellosis associated with exotic pets. J Clin Microbiol 1997;35:2786–2790.

APPENDIX

SUPPLY SOURCES OF SMALL ANIMAL HANDLING AND RESTRAINT EQUIPMENT

Dog and Cat Equipment

Campbell Pet Company
P.O. Box 122
Brush Prairie, WA 98606
800–228–6364
Campbellpet.com
Leashes, collars, muzzles, nets, snares, squeeze cages, capture poles, gloves, stretchers, tongs, cat bags

Jorgensen Labs
1450 Van Buren Ave.
Loveland, Colorado 80538
800–525–5614
jorvet.com
Muzzles, cat bags, restraint collars, gloves

Jeffers Pet
310 W. Saunders Road
Dothan, AL 36301
800–533–3377
jefferspet.com
Dog and cat collars, leads, harnesses, crates, muzzles

Lomir Inc.
213 West Main Street
Malone, NY 12953
877–425–3604
Lomir.com
Restraint jackets, collars, enrichment products, restraint slings, gloves
Patterson Veterinary
822 7th St.
Greeley, CO 80631
800–225–7911

Pattersonvet.com
Muzzles, gloves, restraint collars, pet carriers, cat bags, stretchers, gurneys
Shor-Line
511 Osage Avenue
Kansas City, KS 66105
800–444–1579
Shor-line.com
Cages, kennels, tables

Other Small Mammal Equipment

Conduct Science
6 Liberty Square #2321

Boston, MA 02109
888–267–4324
Conductscience.com
Rodent and rabbit restrainers and cages

Avantor
Radnor Corporate Center
Bldg One, Suite 200
100 Matsonford Road
Radnor, PA 19087
888–897–5463
us.vwr.com
Rodent and rabbit restrainers and jackets

Kent Scientific Corp.
1116 Litchfield St.
Torrington, CT 06790
888–572–8887
Kentscientific.com
Rodent restrainers

Lomir Inc.
213 West Main Street
Malone, NY 12953
877–425–3604
Lomir.com
Restraint jackets, collars, enrichment products, restraint slings, gloves

Companion Bird Equipment

Veterinary Specialty Products
10504 W. 79th St.
Shawnee, KS 66214
800–362–8138
vetspecialtyproducts.com
Companion bird restraints

Reptile Equipment

Tomahawk Live Trap
PO Box 155
Hazelhurst, WI 54531
800–272–8727
Livetrap.com
Gloves; cages; control poles; tongs; shields; nets; bite sticks; snares; leases; crates; snake poles, tubes, cages, and bags

Midwest Tongs, Inc
14505 S. Harris Rd
Greenwood, MO 64034
816–537–4444
Tongs.com
Snake-handling tongs, hooks, bags

Multiple Species Recommended Readings for Small Animal Handling

1. Ackerman N, Aspinall, V. Aspinall's Complete Textbook of Veterinary Nursing, 3rd ed. Elsevier, 2016.

2. Anderson RS, Edney ATB. Practical Animal Handling. Pergamon Press, Oxford, England, 1991.

3. Angulo FJ, Glaser CA, Juranek DD, et al. Caring for pets of immunocompromised persons. J Am Vet Med Assoc 1994;205:1711–1718.

4. Ballard B, Rockett J. Restraint and Handling for Veterinary Technicians and Assistants. Delmar, Clifton Park, NY, 2009.

5. Ballard B, Cheek R. Exotic Animal Medicine for the Veterinary Technician, 3rd ed. Wiley-Blackwell, Ames, IA, 2017.

6. Baker WS, Gray GC. A review of published reports regarding zoonotic pathogen infection in veterinarians. J Am Vet Med Assoc 2009;234:1271–1278.

7. Bassett JM, Thomas J. McCurnin's Clinical Textbook for Veterinary Technicians, 9th ed. Elsevier, St. Louis, MO, 2017.

8. Boyle JE. Crow & Walshaw's Manual of Clinical Procedures in Dogs, Cats, Rabbits, & Rodents, 4th ed. Wiley-Blackwell, Ames, IA, 2016.

9. Campbell KL, Campbell JR. Companion Animals, 2nd ed. Pearson Education, Inc. Upper Saddle River, NJ, 2009.

10. Centers for Disease Control and Prevention: Injury Prevention & Control. www.cdc.gov/injury

11. Colville JL, Berryhill DL. Handbook of Zoonoses: Identification and Prevention. Mosby Inc, St. Louis, MO, 2007.

12. Fowler ME. Restraint and Handling of Wild and Domestic Animals, 3rd ed. Wiley-Blackwell, Ames, IA, 2008.

13. Grandin T. Improving Animal Welfare: A Practical Approach, 2nd ed. CAB International, Cambridge, MA 2015.

14. Grandin T, Johnson C. Animals in Translation: Using the Mysteries of Autism to Decode Animal Behavior. Scribner. New York, NY, 2005.

15. Grandin, T, Johnson C. Animals Make Us Human: Creating the Best Life for Animals. First Mariner Books, New York, NY, 2010.

16. Herron ME, Shreyer T. The pet-friendly veterinary practice: a guide for practitioners. Vet Clin Small Anim 2014;44:451–481.

17. Leahy JR, Barrow P. Restraint of Animals, 2nd ed. Cornell Campus Book Store, Ithaca NY, 1953.

18. National Association of State Public Health Veterinarians Animal Contact Compendium Committee 2013. Compendium of Measures to Prevent Disease Associated with Animals in Public Settings. J Am Vet Med Assoc 2013;243:1270–1288.

19. National Association of State Public Health Veterinarians. Compendium of Standard Precautions for Zoonotic Disease Prevention in Veterinary Personnel. www.nasphv.org/Documents/VeterinaryStandardPrecautions.pdf, 2015.

20. National Center for Infectious Diseases. Healthy Pets Healthy People. www.cdc.gov/healthypets/child.htm, October 2016.

21. Pickering LK, Marano N, Bocchini JA, et al. Exposure to nontraditional pets at home and to animals in public settings: Risks to children. Pediatrics 2008;122:876–886.

22. Price EO. Principles & Applications of Domestic Animal Behavior. Cambridge, MA, CAB International, 2008.

23. Romich JA. Understanding Zoonotic Diseases. Thompson Delmar Learning, Clifton Park, NY, 2008.

24. Sheldon CC, Sonsthagen T, Topel JA. Animal Restraint for Veterinary Professionals, 2nd ed. Elsevier, St. Louis, MO, 2016.

25. Taylor SM. Small Animal Clinical Techniques. Saunders, 2nd ed. Elsevier, St. Louis, MO, 2015.

26. Vanhorn B, Clark RW. Veterinary Assisting: Fundamentals & Applications. Delmar Cengage Learning. Clifton Park, NY, 2010.

27. Warren DM. Small Animal Care & Management, 4th ed. Delmar, Cengage Learning, 2015.

28. Yin S. Low Stress Handling, Restraint and Behavior Modification of Dogs & Cats. CattleDog Publishing, Davis, CA, 2009.

INDEX

Note: Page numbers in italics indicate a figure and page numbers in bold indicate a table on the corresponding page.